Programmable
Controllers
Third Edition

Programmable Controllers
Third Edition

Thomas A. Hughes

ISA—The Instrumentation, Systems, and Automation Society

Notice

The information presented in this publication is for the general education of the reader. Because neither the author nor the publisher have any control over the use of the information by the reader, both the author and the publisher disclaim any and all liability of any kind arising out of such use. The reader is expected to exercise sound professional judgment in using any of the information presented in a particular application.

Additionally, neither the author nor the publisher have investigated or considered the affect of any patents on the ability of the reader to use any of the information in a particular application. The reader is responsible for reviewing any possible patents that may affect any particular use of the information presented.

Any references to commercial products in the work are cited as examples only. Neither the author nor the publisher endorse any referenced commercial product. Any trademarks or tradenames referenced belong to the respective owner of the mark or name. Neither the author nor the publisher make any representation regarding the availability of any referenced commercial product at any time. The manufacturer's instructions on use of any commercial product must be followed at all times, even if in conflict with the information in this publication.

ISA
67 Alexander Drive
P.O. Box 12277
Research Triangle Park, NC 27709

Library of Congress Cataloging-in-Publication Data

Hughes, Thomas A.
   Programmable controllers / Thomas A. Hughes.--3rd ed.
      p. cm.
   ISBN 1-55617-729-1
   1. Programmable controllers. I. Title.
TJ223.P76 H84 2000
629.8'9--dc21
                                      00-011736

ISA Resources for Measurement and Control Series (RMC)

- *Measurement and Control Basics, 2nd Edition (1995)*
- *Industrial Level, Pressure, and Density Measurement (1995)*
- *Industrial Flow Measurement (1990)*
- *Programmable Controllers, 3rd Edition (2001)*
- *Control Systems Documentation: Applying Symbols and Identification (1993)*
- *Industrial Data Communications: Fundamentals and Applications, 2nd Edition (1997)*
- *Real-Time Control Networks (1993)*
- *Automation Systems for Control and Data Acquisition (1992)*
- *Control Systems Safety Evaluation and Reliability, 2nd Edition(1998)*

THIS BOOK IS DEDICATED TO

my wife Ellen, my daughter Audrey, and the rest of my family,
for their love and encouragement over the years.

CONTENTS

PREFACE

Since 1989, this book has been used both as a textbook for Programmable Logic Controller courses and for self-study by thousands of professionals. This applications-based book provides a clear and concise presentation of the fundamental principles of programmable controllers for process and machine control. This third edition covers all phases of programmable controller applications from design and programming to installation, maintenance, and start-up. Coverage of all five standard PLC programming languages: Ladder Diagram, Function Block Diagram, Sequential Function Chart, Statement List, and Structured Text has been increased in this third edition and numereous programming applications and examples have been added to more clearly explain each programming language.

The text provides a complete and comprehensive presentation on the design and programming of programmable controller-based control applications. The material also includes chapters on binary logic fundamentals, electrical and electronic principles, input and output systems, memory and addressing, programming languages, and data communication. The final chapter covers design, installation, and maintenance of programmable controllers in detail.

A chapter has added to increase coverage of PLC languages and programming. All the chapters have been supplemented with new or improved example problems and exercises. Most of the illustrations in the book have been revised and improved. Answers to all the exercises have been added at the end of the book to assist students and instructors.

I would like to express my appreciation my wife Ellen for the long hours spent reviewing all three editions. I would also like to thank the technical reviwers for making numerous constructive comments that improved the overall presentation of this third edition.

ABOUT THE AUTHOR

Thomas A. Hughes, a Senior Member of ISA has 30 years of experience in the design and installation of instrumentation and control systems, including 20 years in the management of instrumentation and control projects for the process and nuclear industries. He is the author of two books: *Measurement and Control Basics, 2nd Edition, (1995)* and *Programmable Controllers, 3rd Edition, (2000)*, both published by ISA.

Mr. Hughes received a B. S. in engineering physics from the University of Colorado, and a M.S. in control systems engineering from Colorado State University. He holds professional engineering licenses in the states of Colorado and Alaska, and has held engineering and management positions with Dow Chemical, Rockwell International, EG&G Rocky Flats, Topro Systems Integration, and the International Atomic Energy Agency. Mr. Hughes has taught numerous courses in electronics, mathematics, and instrumentation systems at the college level and in industry. He is currently the Principal Consultant with Nova Systems Engineering Services in Arvada, Colorado.

1

Introduction to Programmable Controllers

Introduction

Programmable controllers were originally designed to replace relay-based control systems and solid-state, hard-wired logic control panels. However, the modern programmable controller's system is far more complex and powerful.

The most basic function performed by programmable controllers is to examine the status of inputs and, in response, control some process or machines through outputs. The logical combination of inputs to produce an output or outputs is called *control logic*. Several logic combinations are usually required to carry out a control plan or program. This control plan is stored in memory using a programming device that inputs the program into the system. The processor (usually a high-speed microprocessor) periodically scans the control plan in memory in a predetermined sequential order. The amount of time required to examine the inputs and outputs, perform the control logic, and execute the outputs is called the *scan time*.

Figure 1-1 shows a simplified block diagram of a programmable controller. In this diagram, a level switch and panel-mounted pushbutton are wired to input circuits, and the output circuits are connected to an electric solenoid valve and a panel-mounted indicator light. The output devices are controlled by the control program in the logic unit.

Figure 1-1 shows a typical configuration of the early programmable controller applications, which were intended to replace relay or hard-wired logic control systems. The input circuits are used to convert the various field voltages and currents into the low voltage signals (normally

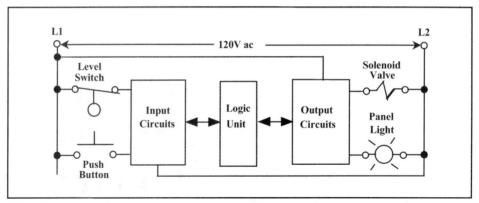

Figure 1-1. Simplified diagram for a programmable controller system.

0 to 5 volts direct current [vdc]) used by the logic unit. The output circuits convert the logic signals to a level that will drive the field devices. For example, in Figure 1-1, 120-volt alternating current (vac) power is connected to the field input devices, so the input circuits are used to convert the 120 vac into the 0- to 5-volt logic signals used by the control unit.

Brief History of PLCs

In 1968, a major automobile manufacturer wrote a design specification for the first programmable controller. The primary goal was to eliminate the high cost associated with the frequent replacement of inflexible relay-based control systems. The specification also called for a solid-state industrial computer that could be easily programmed by maintenance technicians and plant engineers. It was hoped that the programmable controller would reduce production downtime and provide expandability for future production improvements and changes. In response to this design specification, several manufacturers developed computer-based control devices called *programmable controllers*.

The first programmable controller was installed in 1969, and it proved to be a vast improvement over relay-based control systems. The controllers were easy to install and program, they used less plant floor space, and they were more reliable than relay-based control systems. The initial programmable controller not only met the automobile manufacturer's production needs, but further design improvements in later models led to widespread use of programmable controllers in other industries.

Two main factors in the initial design of the programmable controllers probably led to their success. First, highly reliable solid-state components were used, and the electronic circuits were designed for the harsh industrial environment. The input/output (I/O) circuits were designed and built to withstand electrical noise, moisture, oil, and the high temperatures encountered in industry. The second important factor was that the programming language that was initially selected was based on standard electrical ladder logic design. Some earlier computer system applications had failed because plant technicians and engineers could not be trained easily in standard computer software. However, most were already trained in relay ladder logic design, so they could quickly learn programming in a language that was based on the familiar relay ladder diagrams.

When microprocessors were introduced in 1974 and 1975, the basic capabilities of programmable controllers were greatly expanded and improved. They were able to perform sophisticated math and data manipulation functions, which greatly increased the use of programmable controllers in more complex control applications.

In the late 1970s, improved communication components and circuits made it possible to place programmable controllers thousands of feet from the equipment they controlled. Several programmable controllers could now exchange data and thus more effectively control processes and machines. Also, microprocessor-based input and output modules allowed programmable controller systems to evolve into the analog control world.

Programmable controllers are found in thousands of industrial applications. They are used to control chemical, petrochemical, food, pharmaceutical, wastewater treatment, water treatment, nuclear, natural gas, and mining processes. They are found in material transfer and storage systems that transport and store both the raw materials and the finished products. They are used with robots to perform hazardous industrial operations, thus promoting safer operations. Programmable controllers are used in conjunction with other computers to perform process and machine data collection and reporting functions, including statistical process control, quality assurance, and online diagnostics. They are utilized in energy management systems to reduce costs and to improve the environmental control of industrial facilities and office buildings.

The introduction of the personal computer (PC) in the early 1980s greatly increased the power and utility of the programmable controller system in process and machine control. Because personal computers were inexpensive they were used extensively as programming devices and operator interface control stations. The development of low-cost graphical control software packages for PCs has led to the extensive use of *graphical user interfaces* (GUIs) in programmable controller applications.

Because personal computers are used widely both in control and business applications, the abbreviation *PC* is generally reserved for personal computers, and the abbreviation *PLC* is used for programmable controllers or programmable logic controllers. Thus, the abbreviation *PLCs* will be used in this book to represent programmable controllers.

Basic Components of PLC Systems

Regardless of size, cost, or complexity, all programmable controllers share the same basic components and functional characteristics. A programmable controller will always consist of a processor, a memory unit, an input/output system, a programming language, a programming device, and a power supply. A block diagram of a typical PLC system is shown in Figure 1-2.

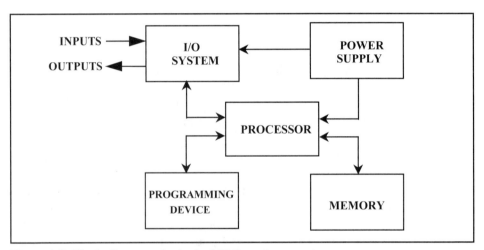

Figure 1-2. Block diagram of a typical PLC system.

The Processor

The processor consists of one or more standard or custom microprocessors and other integrated circuits that perform the logic, control, and memory functions of the PLC system. The processor reads the inputs, executes logic as determined by the application program, performs calculations, and controls the outputs accordingly.

The processor controls the operating cycle or processor scan. This operating cycle consists of a series of operations performed sequentially and repeatedly. A typical PLC processor operating cycle is shown in Figure 1-3.

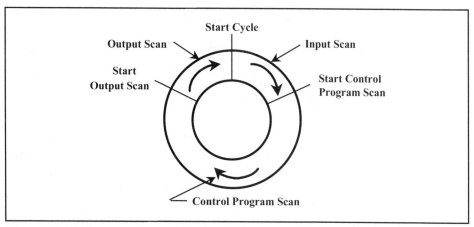

Figure 1-3. PLC processor operating cycle.

During the *input scan*, the PLC examines the external input devices to see if a signal is present or absent, that is, if the input devices are in an ON or OFF state. The status of these inputs is temporarily stored in an input image table or memory file. During the *program scan*, the processor scans the instructions in the control program, uses the input status from the input image file, and determines if an output will or will not be energized. The resulting status of the outputs is written to the output image table or memory file. Based on the data in the output image table, the PLC energizes or deenergizes its associated output circuits, which control external devices. This operating cycle typically takes 1 to 25 milliseconds (thousandths of a second). The input and output scans are normally very short relative to the time required for the program scan.

Memory

Memory is used to store the control program for the PLC system; it is usually located in the same housing as the central processing unit (CPU). The information stored in memory determines how the input and output data will be processed.

Memory stores individual pieces of data called *bits*. A bit has two states: 1 or 0. Memory units are mounted on circuit boards and are usually specified in thousands or "K" increments, where 1K is 1,024 words (i.e., $2^{10} = 1,024$) of storage space. Programmable controller memory capacity may vary from less than 1,000 words to over 64,000 words (64K words) depending on the brand of programmable controller. The complexity of the control plan will determine the amount of memory required.

Although there are several different types of computer memory, they can always be classified as either *volatile* or *nonvolatile*. Volatile memory will lose its programmed contents if all operating power is lost or removed. Volatile memory is easily altered and is quite suitable for most programming applications when they are supported by battery backup or a recorded copy of the program. Nonvolatile memory will retain its data and program even if there is a complete loss of operating power. It does not require a backup system.

The most common form of volatile memory is *random access memory,* or RAM. RAM is relatively fast and provides an easy means to create and store application programs. If normal power is disrupted, PLCs with RAM use battery or capacitor backups to prevent program loss.

Electrically erasable programmable read-only memory (EEPROM) is a nonvolatile memory that is programmed using application software, which runs on a personal computer or through a micro-PLC handheld programmer.

The user can access two areas of memory in the PLC system: *program files* and *data files*. Program files store the control application program, subroutine files, and the error file. Data files store data associated with the control program, such as input/output status bits, counter and timer preset and accumulated values, and other stored constants or variables. Together, these two general memory areas are called *user* or *application memory.* The processor also has an executive or system memory that directs and performs operational activities such as executing the control program and coordinating input scans and output updates. This process system memory, which is programmed by the PLC manufacturer, cannot be accessed or changed by the user.

I/O System

The I/O system provides the physical connection between the process equipment and the microprocessor. This system uses various input circuits or modules to sense and measure the physical quantities of the process, such as motion, level, temperature, pressure, flow, and position. In response to the status sensed or the values measured, the processor controls various output modules. These modules drive field devices such as valves, motors, pumps, and alarms to exercise control over a machine or a process.

Input Types

The inputs from field instruments or sensors supply the data and information the processor needs to make the logical decisions required to control a given process or machine. These input signals come from devices as varied as pushbuttons, hand switches, thermocouples, strain gauges, and so on. The signals are connected to input modules to filter and condition the signal so the processor can use it.

Output Types

The outputs from the programmable logic controller energize or deenergize control devices to regulate processes or machines. These output signals are control voltages from the output circuits, and they are generally not high-power signals. For example, an output module sends a control signal that energizes the coil in a motor starter. The energized coil closes the power contacts of the starter. These contacts then close to start the motor. The output modules are usually not directly connected to the power circuit but rather to devices such as the motor starter and heater contactors that apply high-power (greater than 10 amps) signals to the final control devices.

I/O Structure

PLCs are classified as micro, small, medium, and large mainly based on the I/O count. Micro-PLCs generally have an I/O count of 32 or less, small PLCs have less than 256 I/O points, medium-sized PLCs have an I/O count of less than 1,024, and large PLCs have an I/O count greater than 1,024. Micro-PLCs are self-contained units comprised of the processor, the power supply, and I/O. Because they are self-contained, micro-PLCs are also called *packaged controllers*. A modular PLC is one that has separate components or modules.

A packaged controller offers the advantages of being smaller, costing less, and being easy to install. A typical wiring diagram for a micro-PLC is shown in Figure 1-4. It shows an Allen-Bradley Micro-1000 PLC with nine inputs and five outputs. The unit is powered with 120 vac through an internal power supply that operates the internal I/O circuits and the built-in microprocessor. This power supply also generates 24 volts direct current (vdc) for the field input switches and contacts.

In medium and large PLC systems the I/O modules are normally installed or plugged into a slot in a "universal" modular housing. The term *universal* in this context means that any module can be inserted into any I/O slot in the housing. Modular I/O housings are also normally designed so the I/O modules can be removed without turning off the ac power or removing the field wiring.

Figure 1-5 shows some typical configurations for I/O modular housings. The backplane of the housings into which the modules are plugged has a printed circuit card that contains the parallel communications bus to the processor. It also contains the dc voltages for operating the digital and

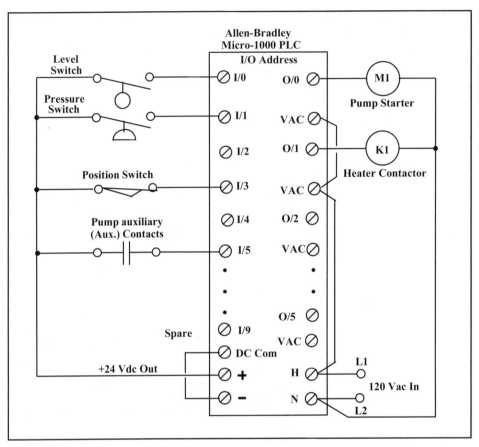

Figure 1-4. Typical micro-PLC wiring diagram.

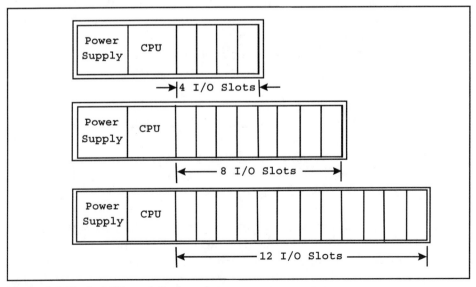

Figure 1-5. Typical I/O modular housings.

analog circuits in the I/O modules. These I/O housings can be mounted in a control panel or on a subpanel in an enclosure. The housings are designed to protect the I/O module circuits from dirt, dust, electrical noise, and mechanical vibration.

The backplane of the I/O chassis has sockets for each module. They provide the power and data communications connection to the processor for each module.

Discrete Inputs/Outputs

Discrete is the most common class of input/output in a programmable controller system. This type of interface module connects field devices that have two discrete states, such as ON/OFF or OPEN/CLOSED, to the processor. Each discrete I/O module is designed to be activated by a field-supplied voltage signal, such as +5 vdc, +24 vdc, 120 vac, or 220 vac.

In a *discrete input (DI)* module, if an input switch is closed an electronic circuit in the input module senses the supplied voltage. To indicate the status of that device it then converts the supplied voltage into a logic-level signal that is acceptable to the processor. A logic 1 indicates "ON" or "CLOSED," and a logic 0 indicates "OFF" or "OPENED" for a field input device or switch. A typical discrete input module is shown in Figure 1-6.

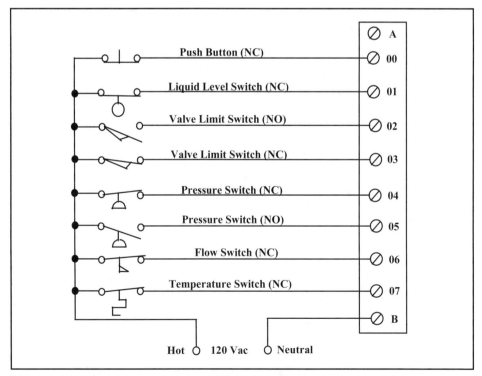

Figure 1-6. Typical discrete input module wiring diagram.

Most input modules will contain a light-emitting diode (LED) to indicate the status of each input.

In a *discrete output (DO)* module, the output interface circuit switches the supplied control voltage that will energize or deenergize the field device. If an output is turned ON through the control program, the interface circuit switches the supplied control voltage to activate the referenced (addressed) output device.

Figure 1-7 shows a wiring diagram for a typical discrete output module. It can be thought of as a simple switch through which power can be provided to control the output device. During normal operation, the processor sends the output state that was determined by the logic program to the output module. The module then switches the power to the field device.

A fuse is normally provided in the output circuit of the module to prevent excessive current from damaging the wiring to the field device. If the fuse is not provided, it should be included in the system design.

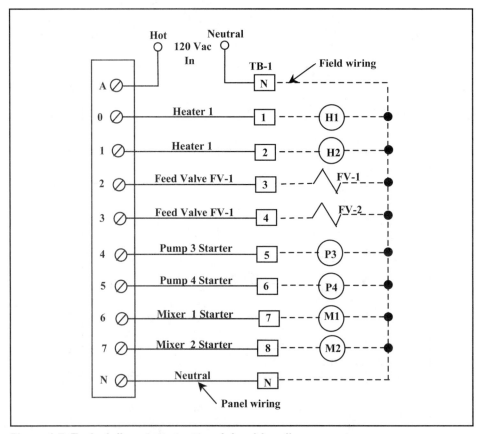

Figure 1-7. Typical discrete output module wiring diagram.

Analog I/O Modules

The analog I/O modules make it possible to monitor and control analog voltages and currents, which are compatible with many sensors, motor drives, and process instruments. By using analog I/O, it is possible to measure or control most process variables with appropriate interfacing.

Analog I/O interfaces are generally available for several standard unipolar (single polarity) and bipolar (negative and positive polarity) ratings. In most cases, a single input or output interface can accommodate two or more different ratings and can satisfy either a current or a voltage requirement. The different ratings can be selected via either hardware (i.e., switches or jumpers) or software.

Digital I/O Modules

Digital I/O modules are similar to discrete I/O modules in that they process discrete ON/OFF signals. However, the main difference is that discrete I/O interfaces require only a single bit to read an input or control an output. On the other hand, digital I/O modules process a group of discrete bits in parallel or serial form.

Typical devices that interface with digital input modules are binary encoders, bar code readers, and thumbwheel switches. LED displays and intelligent display panels are just two of the instruments that are driven by digital output modules.

Special-Purpose Modules

The discrete and analog I/O modules will normally cover about 80 percent of the input and output signals encountered in programmable controller applications. However, to process certain types of signals or data efficiently, the programmable controller system will require special-purpose modules. These special interfaces include those that condition input signals, such as thermocouple modules, pulse counters, or other signals that cannot be interfaced using standard I/O modules. Special-purpose I/O modules may also use an onboard microprocessor to add intelligence to the interface. These intelligent modules can perform complete processing functions independent of the CPU and the control program scan.

Another important class of special-purpose I/O modules are communication modules that communicate with distributed control systems (DCSs), other PLC networks, plant computers, or other intelligent devices.

Programming Languages

The programming language allows the user to communicate with the programmable controller via a programming device. Programmable controller manufacturers use several different programming languages, but they all use instructions to convey a basic control plan to the system.

A control plan or program is defined as a set of instructions arranged in a logical sequence to control the actions of a process or machine. For example, the program might direct the programmable controller to turn a motor starter on when a pushbutton is depressed. It might at the same time direct the programmable controller to turn on a control panel-mounted run light when the motor starter auxiliary contacts are closed.

A program is written by combining instructions in a certain order. Rules govern the way in which instructions are combined as well as the actual form of the instructions. These rules and instructions combine to form a language. The three most common types of languages encountered in programmable controller systems are as follows:

1.   Ladder logic (LAD)

2.   Statement list (STL)

3.   Function block diagram (FBD)

Figure 1-8 shows a simple logic function implemented using each of the three common types of PLC languages. The logic function shown is an **AND** function, that is, if pushbutton 1 (PB_1) is closed *and* pushbutton 2 (PB_2) is closed then the *GO_Light* is on.

Ladder Logic (LAD)

Ladder logic (LAD) is the most common language used in PLC applications. The reason for this is relatively simple. The original programmable controllers were designed to replace electrical relay-based control systems. Those systems were designed by technicians and engineers using a symbolic language called *ladder diagrams*. The ladder diagram consists of a series of symbols interconnected by lines to indicate the flow of current through the various devices. The ladder drawing consists of basically two things. The first is the power source, which forms the sides of the ladder (rails), and the second is the current that flows through the various logic input devices that form the rungs of the ladder. If there is electrical current flow through the relay contacts in a rung, the output relay coil will be turned on. This is termed *power flow* in the ladder rung.

In electrical design, the ladder diagram is intended to show only the circuitry necessary for the basic operation of the control system. Another

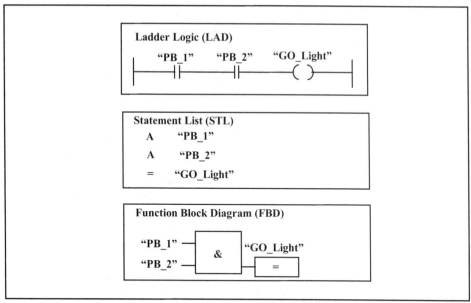

Figure 1-8. A sample logic function using three PLC languages.

diagram, called the *wiring diagram*, is used to show the physical connection of the control devices. The discrete I/O module diagrams shown earlier are examples of wiring diagrams. A typical electrical ladder diagram is shown in Figure l-9. In this diagram, a pushbutton (PBl) is used to energize a pump start control relay (CRl) if the level in a liquid storage tank is not high. Each device has a special symbol assigned to it so the diagram can be read easier and faster.

The same control application can be implemented using a PLC ladder logic (LAD) program, as shown in Figure l-10. The two diagrams are read

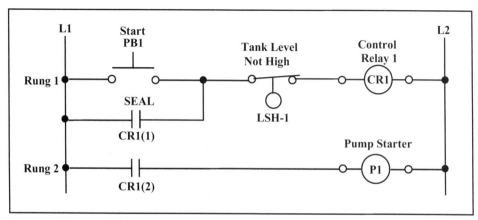

Figure 1-9. Typical electrical ladder diagram.

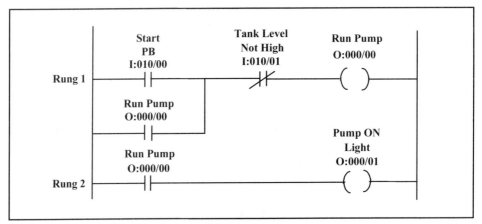

Figure 1-10. Typical ladder diagram (LAD) program.

in the same way, from left to right, with the logic input conditions on the left and the logical outputs on the right. In the case of electrical diagrams, there must be electrical continuity to energize the output devices. For programmable controller ladder programs, there must be logic continuity to energize the outputs.

In ladder programs, three basic instructions are used to form the program. The first symbol is similar to the *normally open* (NO) relay contacts used in electrical ladder diagrams. This instruction uses the same NO symbol in ladder programs. It instructs the processor to examine its assigned bit location in memory. If the bit is ON (logic l), the instruction is TRUE and there is logic continuity through the instruction on the ladder rung. If the bit is OFF (logic 0), there is no logic continuity through the instruction on the rung.

The second important instruction is similar to the *normally closed* (NC) contact from electrical ladder diagrams. It is called the *examine off* instruction. Unlike the *examine on* instruction, it directs the processor to examine the bit for logical 0 or the OFF condition. If the bit is OFF, the instruction is TRUE and there is logic continuity through the instruction. If the bit is ON, the normally closed instruction is FALSE and there is no logic continuity.

The third instruction is the *output coil* instruction. This instruction is similar to the relay coil in electrical ladder diagrams. It directs the processor to set a certain location in memory to ON or l if there is logic continuity in any logic path preceding it. If there are no complete logic continuity paths in the ladder rung, the processor sets the output coil instruction bit to 0 or OFF.

In Figure 1-10, the reference addresses for the logic bits are given by the letters *I* and *O* followed by a five-digit number above the instructions. The letter *I* before the five-digit number indicates an input bit, and the letter *O* before the five-digit number indicates an output bit. The reference address indicates where in the memory the logic operation will take place.

In the ladder logic program shown in Figure 1-10, the *examine on* instruction for the start pushbutton (PB) directs the processor to see if the reference address I:010/00 is ON. In the same way, the *examine on* instruction for the *Tank Level Not High* input instructs the processor to see if the reference address I:010/01 is OFF. If there is logic continuity through both instructions, the output coil instruction at address O: 000/00 is turned ON. Logic continuity from the left side to the right end of a rung is called *logical power flow* in PLC programming.

This same output bit is then used to "seal in" the *start pushbutton* instruction. It also turns on the energized instruction bit O:000/01 to turn on the pump run light.

Statement List (STL)

Statement list (STL) is a textual programming language that you can use to create the code for a PLC control program. Its syntax for statements is similar to microprocessor assembly language and consists of instructions followed by addresses on which the instructions act. The STL language contains a comprehensive range of instructions for creating a complete user program. For example, in the Siemens S7 programming software package, there are over 130 different basic STL instructions and a wide range of addresses available depending on the PLC model you use.

STL instruction statements have two basic structures: one statement comprised of an instruction alone (for example, NOT) and another in which the statement is comprised of both an instruction and an address. The most common structure is for the statement to have an instruction and an address. The address of an instruction statement indicates a constant or the location where the instruction finds a value on which to perform an operation.

The *Boolean bit logic instructions* are the most basic type of STL instructions. These instructions perform logic operations on single bits in PLC memory. The basic bit logic instructions are AND (A) and its negated form AND NOT (AN), OR (O), and EXCLUSIVE OR (OR) and its negated form, EXCLUSIVE OR NOT (XN). These instructions check the signal state of a bit address to establish whether the bit is activated (1) or not activated (0).

Bit logic instructions are also called *relay logic instructions* since they can execute commands that can replace a relay logic circuit. Figure 1-11 is an example of AND logic operation. The STL program is listed on the left side, and the relay logic circuit is shown on the right side for comparison. In this example, the statement list program uses an AND instruction (A) to program two normally open (NO) contacts in series. Only when the signal state of both the normally open contacts is 1 can the state of output *Q4.0* be 1 and the coil be energized.

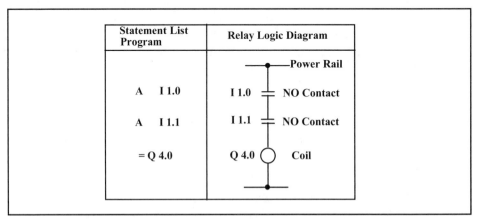

Figure 1-11. Comparison of STL program and relay logic circuit.

Function Block Diagram (FBD)

The *function block diagram (FBD)* is a graphical programming language. It allows the programmer to build complex control procedures by taking existing functions from the FBD library and wiring them in a graphic diagram area. An FBD describes a relationship or function between input and output variables. A function is described as a set of elementary function blocks, as shown in Figure 1-12. Input and output variables are connected to blocks by connection lines.

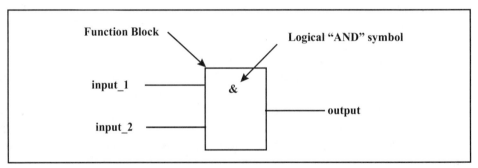

Figure 1-12. Typical elementary function block.

You can build an entire function operated by an FBD program with standard elementary function blocks from the FBD library. Each elementary function block has a fixed number of input connection points and a fixed number of output connection points. For example, the Boolean **AND** function block shown in Figure 1-12 has two inputs and only one output. The inputs are connected on its left border. The outputs are connected on its right border. An elementary function block performs a single function between its inputs and its outputs. For example, the elementary function block shown in Figure 1-12 performs the Boolean **AND** operation on its two inputs and produces a result at the output. The name of the function to be performed by the block is written in its symbol. In the case of the **AND** function the symbol is *&*.

Programming Devices

Programming devices are used to enter, store, and monitor the programmable controller software. They can be dedicated portable unit systems or personal computer-based systems. The personal computer-based systems normally have these basic components: keyboard and mouse; color graphics display or CRT; personal computer; printer; and communications interface card and cable, as shown in Figure 1-13.

You normally connect the programming devices to the programmable controller system only while programming, starting up, or troubleshooting the control system. Otherwise, the programming device is disconnected from the system.

The programming terminals are normally either a handheld programmer or a personal computer-based system. The handheld programmers are inexpensive and portable; they are normally used to program small

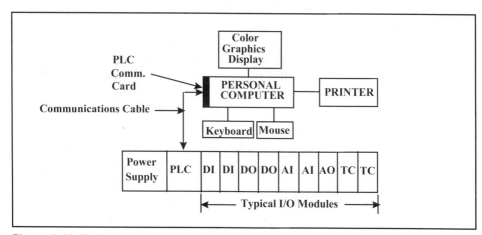

Figure 1-13. Typical programmable controller system.

programmable controllers. Most of these units resemble portable calculators but with larger displays and a somewhat different keyboard. The displays are generally LED (light-emitting diode) or dot matrix LCD (liquid crystal display). The keyboard consists of alphanumeric keys, programming instruction keys, and special-function keys. Even though they are mainly used for writing and editing the control program, the portable programmers are also used for testing, changing, and monitoring the program.

The standard programming terminal is a personal computer like the one shown in Figure 1-13 in which the programming software loaded on the hard drive. These units can perform program editing and storage. They also have added features such as the option to print out programs and connect them to local area networks (LANs). LANs give the programmer or engineer access to any programmable controller in the communications network, so you can monitor and control any device in the network. Normally, laptop PCs are used because they are light and portable and can be easily used in the field while testing, starting up, and modifying the control program.

Power Supply

The power supply converts ac line voltages to dc voltages to power the electronic circuits in a programmable controller system. These power supplies rectify, filter, and regulate voltages and currents to supply the correct amounts of voltage and current to the system. The power supply normally converts 120 vac or 240 vac line voltage into dc voltages such as +5 vdc, −15 vdc, or +15 vdc.

The power supply for a programmable controller system may be integrated with the processor, memory, and I/O modules into a single housing, or it might be a separate unit connected to the system through a cable. As a system expands to include more I/O modules or special-function modules, most programmable controllers require an additional or auxiliary power supply to meet the increased power demand. Programmable controller power supplies are usually designed to eliminate the electrical noise that is present on the ac power and signal lines of industrial plants so it does not introduce errors into the control system. They are also designed to operate properly in the higher-temperature and humidity environments present in most industrial applications.

Graphical User Interface

Two methods are commonly used to provide operators with color process graphics displays in programmable controller-based control systems. The first is to hard-wire the programmable controller I/O modules to a graphics display panel with hardwired lights and digital indicators. This method is cost effective in a small system that will not be changed. It is generally not recommended in larger control systems that will be expanded in the future. The second method is to use an industrial-grade personal computer loaded with process color graphics software. This method has the following advantages: you can easily modify the process display screens for process changes, and the computer can perform other functions such as listing alarms, generating reports, and programmable controller software.

These features are best explained by considering a personal computer-based graphical user interface (GUI) system on a typical process control system, as shown in Figure 1-14. The software for vendor-supplied GUI color displays is normally menu-driven and relatively easy to use. The process display screens are usually based on the process and instrument drawings for the process being controlled.

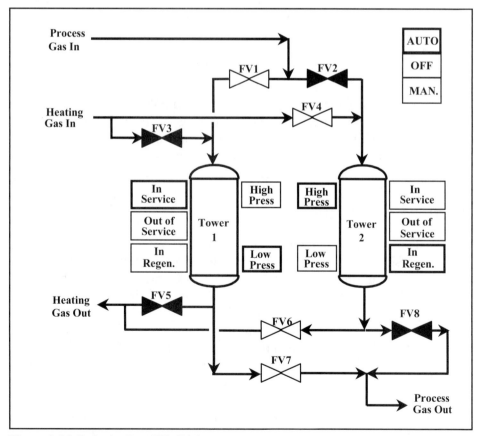

Figure 1-14. Dehydration GUI display.

For example, if we built a process display based on a dehydration process, the computer-generated display would be as shown in Figure 1-14. An advantage of a PC-based GUI is that process data and alarm messages can be displayed on the screen.

In the GUI graphics of Figure 1-14, Tower 1 is shown in service so that its process gas inlet valve FV-1 and process gas outlet valve FV-7 are open and its heating gas inlet valve FV-3 and heating gas outlet valve FV-5 are closed. Tower 2 is shown in regeneration with valves FV-4 and FV-6 opened and valves FV-2 and FV-8 closed. The open valves in Figure 1-14 have no fill color, and the closed valves are solid black. We can also display process conditions on the graphics screen, such as tower pressure high or low, as shown in Figure 1-14.

The process values, such as valves open and closed, are transmitted to the GUI application with special communications interface software called *drivers*. The GUI software is normally configured using a cross-referenced table that lists the PLC input and output addresses that will drive or animate each point or information window on the graphic displays. Normally, a large number of communications drivers are provided with the GUI software so you can interface with the various PLC manufacturer and PLC types. The GUI programmer simply selects the correct PLC manufacturer and the correct PLC model number for the application.

Another important advance in programmable controller is using personal computers to directly replace PLC processors in applications. Generally, the standard PLC input and output modules are controlled by an industrial-grade personal computer. These systems are generally called *SoftLogic* or *SoftPLC* systems.

SoftLogic or SoftPLC

The original hard-logic PLC had the advantage of being designed for harsh industrial environments. However, with recent designs of industrial-grade personal computers this advantage has been lost. One disadvantage of most PLCs is that they use proprietary software. With personal computers, faster and more powerful capabilities are introduced at rapid and regular intervals. Personal computer speeds have historically doubled every eighteen months. This is not true of PLCs, however, because their proprietary architectures require additional engineering so advanced microprocessor technology can be adapted to them.

As a result, it is estimated that hard-logic PLC performance lags personal computer industry advances by eighteen months to two years, and the gap is widening. With PC processing power leapfrogging itself at this same rate, PLCs are having great difficulty keeping up with commodity PCs retailing to consumers for under $1,000. Moreover, after a decade of exponential growth in their capabilities, PCs now come in rugged packages, with ever faster processors, real-time operating systems to handle time-critical operations, larger memory, and a multitude of Windows-based software products. All this comes at ever lower cost because of competitive pricing and high-volume manufacturing.

Thus, the personal computer is starting to replace the PLC processor in small- and medium-sized automation projects. Some PLC manufacturers are actually using PC circuit cards inside their PLCs and calling the system a "SoftPLC." The potential for increased use of personal computers to directly replace proprietary PLCs appears strong.

EXERCISES

1.1   Explain the operation and purpose of the processor in a typical programmable controller system.

1.2   What is the main purpose of the input and output system in a PLC system?

1.3   List some discrete input devices typically found in process industries.

1.4   List some discrete output devices typically encountered in industrial applications.

1.5   List some analog signal values typically found in process applications.

1.6   Explain the difference between volatile and nonvolatile memory.

1.7   List some common applications for personal computers in programmable controller systems.

1.8   What device is most commonly used to program PLCs?

1.9   Discuss the three basic instructions used in ladder logic programs.

1.10   What is power flow in a relay ladder diagram?

1.11   Explain the concept of logical power flow in a ladder diagram program.

1.12   What are the two basic structures used in STL instruction statements?

BIBLIOGRAPHY

1. Allen-Bradley Co., Inc. *Micro Mentor--Understanding and Applying Micro Programmable Controllers* (Allen-Bradley, 1995).

2. Gilbert, R. A., and J. A. Llewellyn. *Programmable Controllers— Practices and Concepts* (Industrial Training Corp., 1985).

3. Jones, C. T., and L. A. Bryan. *Programmable Controllers Concepts and Applications* (International Programmable Controllers, 1983).

4. Plato Computer-Based Training. *Programmable Controller Fundamentals* (Allen- Bradley, 1985).

5. Webb, J. W., and R. A. Reis. *Programmable Logic Controllers— Principles and Applications*, 3d ed. (Prentice-Hall, 1995).

6. Wisnosky, D. E. *SoftLogic: Overcoming Funnel Vision* (Wizdom Controls, 1996).

2

Binary Logic Fundamentals

Introduction

Designing and maintaining programmable controllers in process control applications requires an understanding of binary logic fundamentals. In this chapter, we will discuss the basic concepts of binary signals, numbering systems, binary data codes, binary logic functions, logic function symbols, and relay ladder logic design. Since data in programmable logic controllers is in binary form (0 or 1), we will first discuss binary signals and codes.

Binary Signals and Codes

To use PLCs in process control the measurement and control signals must be encoded into binary form. Binary signals are simply two-state signals (ON/OFF, start/stop, high voltage/low voltage, etc.).

The simplest approach to encoding analog data into a binary-based digital word is provided by the American Standard Code for Information Interchange (ASCII). This method uses a pattern of seven bits (ones and zeros) to represent letters and numbers. Sometimes an extra bit (called a parity bit) is used to check that the correct pattern has been transmitted. For example, the number 1 is represented by 011 0001, and the number 2 is represented by 011 0010. On the other hand, the capital letter *A* is represented by 100 0001 and the capital letter *B* by 100 0010.

Many other methods are used to encode digital numbers in programmable controllers and digital computers. The most common method is the simple binary code, which we will discuss later. First, however, we will briefly cover numbering systems.

Numbering Systems

The most commonly used numbering system in programmable controllers is the binary system, but the octal and hexadecimal numbering systems are also encountered. We will start with a brief review of the decimal numbering system and then discuss each of the other three systems.

Decimal Numbering System

The decimal numbering system is probably in common use because man started to count with his fingers. However, it is not an easy system to implement electronically; a ten-state electronic device would be quite costly and complex. It is much easier and more efficient to use the binary (two-state) numbering system when manipulating numbers using the logic circuits found in computers.

A decimal number, N_{10}, can be written mathematically as follows:

$$N_{10} = d_n R^n + ... + d_2 R^2 + d_1 R^1 + d_0 R^0 \qquad (2\text{-}1)$$

where R is equal to the number of digit symbols used in the system. R is called the *radix* and is equal to 10 in the decimal system. The subscript 10 on the number N in Equation 2-1 indicates that it is a decimal number. However, it is common practice to omit this subscript when writing out decimal numbers. The decimal digits, $d_n ... d_2, d_1, d_0$, can assume the values of 0, 1, 2, 3, 4, 5, 6, 7, 8, or 9 in the decimal numbering system.

For example, the decimal number 1,735 can be written as follows:

$$1735 = 1 \text{x} 10^3 + 7 \text{x} 10^2 + 3 \text{x} 10^1 + 5 \text{x} 10^0$$

$$1735 = 1000 + 700 + 30 + 5$$

When written as 1,735, the powers of ten are implied by positional notation. The value of the decimal number is computed by multiplying each digit by the weight of its position and then summing the result. As we will see, this is true for all numbering systems. The decimal equivalent of any number can be calculated by multiplying the digit by its base raised to the power of the digit's position. The general equation for numbering systems is the following:

$$N_b = Z_n R^n + ... + Z_2 R^2 + Z_1 R^1 + Z_0 R^0 \qquad (2\text{-}2)$$

where Z is the value of the digit, and R is the radix or base of the numbering system.

Binary Numbering System

The binary numbering system has a base of two, and the only allowable digits are 0 or 1. This is the basic numbering system for computers and programmable controllers, which are basically electronic devices that manipulate 0s and 1s to perform math and control functions.

It was easier and more convenient to design digital computers that operate on two entities or numbers than the ten numbers used in the decimal world. Furthermore, most physical elements in the process environment have only two states, such as a pump on or off, a valve open or closed, a switch on or off, and so on.

A binary number follows the same format as a decimal one, that is, the value of a digit is determined by its position in relation to the other digits in a number. In the decimal system, a 1 by itself is worth 1. Placing it to the left of a zero makes the 1 worth 10, and putting it to the left of two zeros makes it worth 100. This simple rule is the foundation of the numbering systems. For example, numbers to be added or subtracted are first arranged so that their place columns line up.

In the decimal system, each position to the left of the decimal point indicates an increasing power of ten. In the binary system, each place to the left signifies an increased power of two, that is, 2^0 is one, 2^1 is two, 2^2 is four, 2^3 is eight, and so on. So, finding the decimal equivalent of a binary number is simply a matter of noting which place columns the binary 1s occupy and then adding up their values. A binary number also uses standard positional notation. The decimal equivalent of a binary number can be found using this equation:

$$N_2 = Z_n 2^n + \ldots + Z_2 2^2 + Z_1 2^1 + Z_0 2^0 \tag{2-3}$$

where the radix or base equals 2 in the binary system, and each binary digit (bit) can only have the value 0 or 1.

The decimal equivalent of the binary number 10101_2 can be found as follows:

$$10101_2 = 1 \times 2^4 + 0 \times 2^3 + 1 \times 2^2 + 0 \times 2^1 + 1 \times 2^0$$

or

$$(1 \times 16) + (0 \times 8) + (1 \times 4) + (0 \times 2) + (1 \times 1) = 21 \text{ (decimal)}$$

EXAMPLE 2-1

Problem: Convert the binary number 101110_2 into its decimal equivalent.

Solution: Using Equation 2-3:

$$N_2 = Z_n2^n + \ldots + Z_22^2 + Z_12^1 + Z_02^0$$

or

$$N_2 = 1 \times 2^5 + 0 \times 2^4 + 1 \times 2^3 + 1 \times 2^2 + 1 \times 2^1 + 0 \times 2^0$$

$$= (1 \times 32) + (0 \times 16) + (1 \times 8) + (1 \times 4) + (1 \times 2) + (0 \times 1)$$

$$= 32 + 0 + 8 + 4 + 2 + 0$$

$$= 46$$

Up to this point, we have discussed only positive binary numbers. Several common methods are used to represent negative binary numbers in programmable controller systems. The first is *signed-magnitude binary*. This method places an extra bit (sign bit) in the left-most position and lets this bit determine whether the number is positive or negative. The number is positive if the sign bit is 0 and negative if the sign bit is 1. For example, in a sixteen-bit machine, suppose we have a twelve-bit binary number, $000000010101_2 = 21_{10}$. To express the positive and negative values we would manipulate the left-most or most significant bit (MSB). So, using the signed magnitude method would yield: $0000000000010101 = +21$ and $1000000000010101 = -21$.

Another common method used to express negative binary numbers is called *two's complement binary*. To complement a number means to change it to a negative number. For example, the binary number 10101 is equal to decimal 21. To get the negative using the two's complement method, you complement each bit and then add one to the least significant bit (LSB).

In the case of the binary number $010101 = 21$, its two's complement would be: $101011 = -21$.

Octal Numbering System

In the binary numbering system substantially more digits are needed to express a number than in the decimal system. For example, $130_{10} = 10000010_2$, so it takes eight or more binary digits to express a decimal number over 127. It is also difficult for people to read and manipulate large numbers without making errors. To reduce errors in binary number manipulations, some computer manufacturers started using the octal numbering system. This system uses the number eight as a base or radix with the eight digits 0, 1, 2, 3, 4, 5, 6, 7. Like all other number systems, each digit in an octal number has a weighted value according to its position.

For example:

$$1301_8 = 1 \times 8^3 + 3 \times 8^2 + 0 \times 8^1 + 1 \times 8^0$$

$$= 1 \times 512 + 3 \times 64 + 0 \times 8 + 1 \times 1$$

$$= 512 + 192 + 0 + 1$$

$$= 705_{10}$$

The octal system is used as a convenient means of writing or manipulating binary numbers in PLC systems. A binary number with a large number of ones and zeros can be represented by an equivalent octal number with fewer digits. As shown in Table 2-1, one octal digit can be used to express three binary digits, so the number is reduced by a factor of three.

Table 2-1. Binary and Octal Equivalent Numbers

Binary	Octal
000	0
001	1
010	2
011	3
100	4
101	5
110	6
111	7

For example, the binary number 11010101010_2 can be converted into an octal number by grouping binary bits in groups of three starting with the least significant bit, as follows:

$$11\ 010\ 101\ 010 = 3252_8$$

EXAMPLE 2-2

Problem: Represent the binary number **101011001101111_2** in octal.

Solution: To convert from a binary to an octal number, we simply divide the binary number into groups of three bits, starting with the least significant bit. Then we use Table 2-1 to convert the three-bit groups into their octal equivalent.

To solve, we place the binary number into groups of three:

$$101 \ 011 \ 001 \ 101 \ 111_2$$

Since $101_2 = 5_8$, $011_2 = 3_8$, $001_2 = 1_8$, $101_2 = 5_8$, and $111_2 = 7_8$, we obtain

$101011001101111_2 = 53157_8$.

We can convert a decimal number into an octal number by successively dividing the decimal number by the octal base number eight. This is best illustrated in the following example.

EXAMPLE 2-3

Problem: Convert the decimal number 370_{10} into an octal number.

Solution: Decimal-to-octal conversion is obtained by successive division by the octal base number 8, as follows:

Division	Quotient	Remainder
370/8	46	2 (LSD)
46/8	5	6
5/8	0	5 (MSD)

Thus, $370_{10} = 562_8$.

Hexadecimal Numbering System

The hexadecimal numbering system provides an even shorter notation than the octal system and is a commonly used numbering system in PLC applications. The hexadecimal system has a base of sixteen, and four binary bits are used to represent a single symbol. The sixteen symbols are 0, 1, 2, 3, 4, 5, 6, 7, 8, 9, A, B, C, D, E, and F. The letters A through F are used to represent the binary strings 1010, 1011, 1100, 1101, 1110, and 1111, which correspond to the decimal numbers ten through fifteen. The hexadecimal digits and their binary equivalents are given in Table 2-2.

Table 2-2. Hexadecimal and Binary Equivalent Numbers

Hexadecimal	Binary	Hexadecimal	Binary
0	0000	9	1001
1	0001	A	1010
2	0010	B	1011
3	0011	C	1100
4	0100	D	1101
5	0101	E	1110
6	0110	F	1111
7	0111	10	10000
8	1000	11	10001

To convert a binary number into a hexadecimal number, we use Table 2-2. For example, the binary number 0110 1111 1000 is 6F8. Again, the hexadecimal numbers follow the standard positional convention $H_n. \ldots H_2, H_1, H_0$, where the positional weights for hexadecimal numbers are powers of sixteen with 1, 16, 256, and 4,096 being the first four decimal values.

To convert from hexadecimal to decimal numbers, we use the following equation:

$$N_{16} = H_n 16^n + \ldots + H_2 16^2 + H_1 16^1 + H_0 16^0 \tag{2-4}$$

where the radix equals sixteen in the hexadecimal system, and each digit can take on the value of zero through nine and the letters $A, B, C, D, E,$ and F.

EXAMPLE 2-4

Problem: Convert the hex number 1FA into its decimal equivalent.

Solution: Using Equation 2-4:

$$N_{16} = H_n 16^n + \ldots + H_2 16^2 + H_1 16^1 + H_0 16^0$$

and since $H_2 = 1 = 1_{10}$, $H_I = F = 15_{10}$, and $H_O = A = 10_{10}$, we obtain:

$$N_{16} = 1 \times 16^2 + 15 \times 16^1 + 10 \times 16^0$$

$$= 256 + 240 + 10 = 506_{10}$$

To convert a decimal number into a hexadecimal number, we use the following procedure:

1.   Divide the decimal number by sixteen and record the quotient and the remainder.

2.   Divide the quotient from the division in Step 1 by sixteen and record the quotient and the remainder.

3.   Repeat Step 2 until the quotient is zero.

4.   The hexadecimal equivalents of the remainders generated by the divisions are the digits of the hexadecimal number, where the first remainder is the least significant digit (LSD) and the last remainder is the most significant digit (MSD).

An example will illustrate the conversion from a decimal into hexadecimal number.

EXAMPLE 2-5

Problem: Convert the decimal number 610_{10} into a hexadecimal number.

Solution:

Division	Quotient	Remainder
610/16	38	2 (LSD)
38/16	2	6
2/16	0	2 (MSD)

Therefore, $610_{10} = 262_{16}$.

It is important to note that the octal and hexadecimal systems are used for human convenience only. The computer system actually converts the octal and hex numbers into binary strings and operates on the binary digits.

Binary Data Codes

Data codes translate information (alpha, numeric, or control characters) into a form that can be transferred electronically and then converted back to its original form. A code's efficiency is a measure of its ability to utilize the maximum capacity of the bits and to recover from error. In the evolution of the various codes, their efficiency at transferring data has steadily increased. A brief discussion of the four commonly used codes follows.

Binary Code

It is possible to represent 2^n different symbols in a purely binary code of n bits. The binary code is a direct conversion of the decimal number into the binary. This is illustrated in Table 2-3.

Table 2-3. Binary to Decimal Code

Decimal	Binary	Decimal	Binary
0	00000	11	01011
1	00001	12	01100
2	00010	13	01101
3	00011	14	01110
4	00100	15	01111
5	00101	16	10000
6	00110	17	10001
7	00111	18	10010
8	01000	19	10011
9	01001	20	10100
10	01010	21	10101

Binary code is the most commonly used code in computers because it is a systematic arrangement of the digits. It is also a weighted code, where each column has a magnitude of 2^n associated with it, and it is easy to translate. In Table 2-3, note that the least significant bit alternates every time, whereas the second least significant bit repeats every two times, the third least significant bit repeats every four times, and so on.

Baudot Code

The Baudot code was the first successful data communications code. It is also known as the International Telegraphic Alphabet #2 (ITA#2). The code was meant primarily for transmitting text. It has only uppercase letters and is used with punched paper tape units on teletypewriters. It uses five consecutive bits and an additional start/stop bit to represent a data character, as shown in Figure 2-1. It transfers asynchronous data at a very slow rate (ten characters per second) using a teletypewriter.

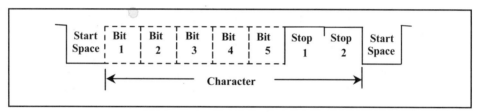

Figure 2-1. Baudot character communication format.

Most early teletypewriters used basically the same circuit as the telegraph with the mechanics of a typewriter. As with the telegraph, the teletypewriter had to have at one end a means for knowing when the other end wanted to transmit, so a mark signal (current) would be sent as a "line idle" signal. Since a mark is the idle condition, the first element or bit of any code would have to be a space (no current). This bit is known as the "start space,"as shown in Figure 2-1. Also, the "current on" (mark) condition needs to exist after the character code pulses have been sent. This enables the receiver device to know when the character is complete and separate this transmission character from the next character to be transmitted. This period of current is know as the "stop mark" and is either 1, 1.42, or 2 elements in duration. The bit time or duration is determined by the teletype's motor speed.

The Baudot or teletypewriter code is given in Table 2-4. There are twenty-six uppercase letters, ten numerals, and various items of punctuation and teletype control. This code uses five bits (two to a fifth power) or thirty-two patterns. However, the code developers used the mechanical shift of the teletypewriter to produce twenty-six patterns out of a possible thirty-two for letters and twenty-six patterns shifted for numbers and punctuation. Only twenty-six were available in either shift because six patterns were the same for both, as shown in Table 2-4. The six common patterns are carriage return, line feed, shift up (figures), shift down (letters), space, and blank (no current).

Table 2-4. Baudot Code

Character Case		Bit Pattern	Character Case		Bit Pattern
Lower	Upper	54321	Lower	Upper	54321
A		00011	Q	1	10111
B	?	11001	R	4	01010
C	:	01110	S	'	00101
D	$	01001	T	5	10000
E	3	00001	U	7	00111
F	!	01101	V	;	11110
G	&	11010	W	2	10011
H	#	10100	X	/	11101
I	8	00110	Y	6	10101
J	Bell	01011	Z	"	10001
K	(	01111	Letters Shift Down		11111
L	)	10010	Figures Shift Up		11011
M	.	11100	Space		00100
N	,	01100	Carriage Return		01000
O	9	11000	Line Feed		00010
P	0	01101	Blank or Null		00000

This binary code is still the most efficient code for narrative text in terms of transmission overhead because it requires very little machine operation or error detection. While this code is no longer widely used, at one time it was the most extensively used binary transmission code.

A disadvantage of the code is that it can represent only the fifty-eight characters shown in Table 2-4. Other limitations of Baudot code are its sequential nature, the high overhead, and the lack of error detection.

BCD Code

As computer and data communications technology improved, more efficient codes were developed. The BCD, or *binary-coded decimal* code, was first used to perform internal numeric calculations within data processing devices. The BCD code is commonly used in programmable controllers to code data to numeric light-emitting diode (LED) displays and from panel-mounted digital thumbwheel units. Its main disadvantages are that it has no alpha characters and no error-checking capability. A listing of the BCD code for decimal numbers from zero to nineteen is given in Table 2-5.

Note that four-bit groups are used to represent the decimal numbers zero through nine. To represent higher numbers, such as ten through nineteen, another four-bit group is used and placed to the left of the first four-bit group.

Table 2-5. BCD Code

Decimal	BCD Code	Decimal	BCD Code
0	0000	10	0001 0000
1	0001	11	0001 0001
2	0010	12	0001 0010
3	0011	13	0001 0011
4	0100	14	0001 0100
5	0101	15	0001 0101
6	0110	16	0001 0110
7	0111	17	0001 0111
8	1000	18	0001 1000
9	1001	19	0001 1001

EXAMPLE 2-6

Problem: Convert the following decimal numbers to BCD code:

    (a) 276,

    (b) 567,

    (c) 719, and

    (d) 4500.

Solution: Using Table 2-5, the decimal numbers can be expressed in BCD code as follows:

    (a) 276 = 0010 0111 0110,

    (b) 567 = 0101 0110 0111,

    (c) 719 = 0111 0001 1001, and

    (d) 4500 = 0100 0101 0000 0000

ASCII Code

The most widely used code is ASCII, which was developed in 1963. This code has seven bits for data (allowing 128 characters), as shown in Table 2-6.

Table 2-6. ASCII Code

Bits	7	0	0	0	0	1	1	1	1
	6	0	0	1	1	0	0	1	1
	5	0	1	0	1	0	1	0	1
4321	HEX	0	1	2	3	4	5	6	7
0000	0	NUL	DLE	SP	0	@	P	`	p
0001	1	SOH	DC1	!	1	A	Q	a	q
0010	2	STX	DC2	"	2	B	R	b	r
0011	3	ETX	DC3	#	3	C	S	c	s
0100	4	EOT	DC4	$	4	D	T	d	t
0101	5	ENQ	NAK	%	5	E	U	e	u
0110	6	ACK	SYN	&	6	F	V	f	v
0111	7	BEL	ETB	'	7	G	W	g	w
1000	8	BS	CAN	(	8	H	X	h	x
1001	9	HT	EM	)	9	I	Y	i	y
1010	A	LF	SUB	*	:	J	Z	j	z
1011	B	VT	ESC	+	;	K	[	k	{
1100	C	FF	FS	'	<	L	\	l		
1101	D	CR	GS	-	=	M	]	m	}
1110	E	SO	RS	.	>	N	^	n	~
1111	F	SI	US	/	?	O	-	o	DEL

The ASCII code can operate synchronously or asynchronously with one or two stop bits. ASCII format has thirty-two control characters. The legend for these control characters is listed in Table 2-7. These control codes are used to indicate, modify, or stop a control function in the transmitter or receiver. Seven of the ASCII control codes are called *format effectors.* They pertain to the control of a printing device. The use of format effectors increases code efficiency and speed by replacing frequently used character combinations with a single code. The format effectors used in the ASCII code are as follows: BS (backspace), HT (horizontal tab), LF (line feed), VT (vertical tab), FF (form feed), CR (carriage return), and SP (space).

EXAMPLE 2-7

Problem: Express the words *PUMP 100 ON* using ASCII code. Use hex notation for brevity.

Solution: Using Table 2-6, we obtain:

MESSAGE:	PUMP 100 ON
ASCII (Hex) String:	50 55 4D 50 20 31 30 30 20 4F 4E

Table 2-7. Legend for ASCII Control Characters

Mnemonic	Meaning	Mnemonic	Meaning
NUL	Null	DLE	Data Link Escape
SOH	Start of Heading	DC1	Device Control 1
STX	State of Text	DC2	Device Control 2
ETX	End of Text	DC3	Device Control 3
EOT	End of Transmission	DC4	Device Control 4
ENQ	Enquiry	NAK	Negative Acknowlege
ACK	Acknowledge	SYN	Synchronous Idle
BEL	Bell	ETB	End of Transmission Block
BS	Backspace	CAN	Cancel
HT	Horizontal Tabulation	EM	End of Medium
LF	Line Feed	SUB	Substitute
VT	Vertical Tabulation	ESC	Escape
FF	Form Feed	FS	File Separator
CR	Carriage Return	GS	Group Separator
SO	Shift Out	RS	Record Separator
SI	Shift In	US	Unit Separator
		DEL	Delete

Binary Logic Functions

In control system applications, the binary numbers 1 and 0 are represented by voltage levels, relay contact status, switch position, and so on. For example, in transistor-transistor logic (TTL) gates, a binary 1 is represented by a voltage signal in the range of 2.4 to 5.0 volts, and a binary 0 is represented by a voltage level between 0 and 0.8 volt. Solid-state electronic circuits are available that you can use to manipulate digital signals to perform a variety of logical functions, such as **NOT, AND, OR, NAND**, and **NOR**. In hardwired electrical logic systems, electrical relays are used to implement the logic functions.

NOT Function

The most basic binary logic function is the **NOT** or inversion function. The **NOT**, or logic inverter, produces an output that is opposite to the input. An inversion bar is drawn over a logic variable to indicate the **NOT** function. For example, if a **NOT** operation is performed on a logic variable A, it is designated by $Z = \overline{A}$. The binary logic truth table for the **NOT** function in Table 2-8 lists the results of the **NOT** function on the input A. In relay-based logic circuits, a normally closed (NC) set of contacts is used to perform the **NOT** function, as shown in Figure 2-2. If the electric relay A is not energized, there is electrical current flow or logic continuity through the normally closed contacts. As a result, that output relay coil is energerized or ON. In other words, if input A is logic 0 or not ON then the output Z is logic 1 or ON. If input A is logic 1 or ON then the normally closed contacts are opened. In addition, there is no current flow or logic flow in the circuit, and relay Z is off and output Z is 0.

Table 2-8. NOT Function Binary Logic Truth Table

Input	Output
A	Z
0	1
1	0

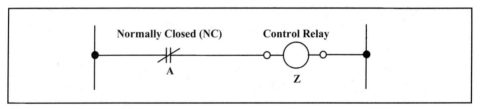

Figure 2-2. NOT function implemented with relay.

OR Function

A logical **OR** function, with two or more inputs and a single output, operates in accordance with the following definition: *The output of an OR function assumes the 1 state if one or more inputs assume the 1 state.*

The inputs to a logic function **OR** gate can be designated by $A, B, \ldots, N$ and the output by Z. It is assumed that the inputs and outputs can take only one of two possible values, either 0 or 1. The logic expression for this function is $Z = A + B + \ldots + N$. A two-input truth table for an **OR** function is given in Table 2-9 for a two-input **OR** function.

Table 2-9. Two-Input OR Function Truth Table

Input A B	Output Z
0 0	0
0 1	1
1 0	1
1 1	1

An example of **OR** logic in process control would be as follows: If the water level in a hot water heater is low or the temperature in the tank is too high, a logic system can be designed to turn off the heater in the system using logic circuits or relays. Figure 2-3 shows this application using relays to perform the logic function.

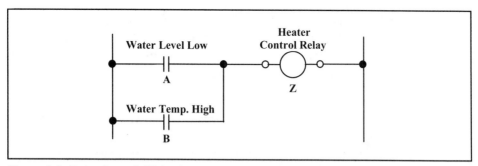

Figure 2-3. Relay-based OR logic control application.

You can easily verify the following logic identities for **OR** functions by using the two-input truth table for the **OR** function given in Table 2-9:

$$A + B + C = (A + B) + C = A + (B + C) \tag{2-5}$$

$$A + B = B + A \qquad (2\text{-}6)$$

$$A + A = A \qquad (2\text{-}7)$$

$$A + 1 = 1 \qquad (2\text{-}8)$$

$$A + 0 = A \qquad (2\text{-}9)$$

Remember that A, B, and C can take on only the value of 0 or 1.

AND Function

An **AND** function has two or more inputs and a single output, and it operates in accordance with the following rule: *The output of an AND gate assumes the 1 state if and only if all the inputs assume the 1 state.* The general equation for the **AND** function is given by $ABC \ldots N = Z$. A two-input **AND** function truth table is given in Table 2-10.

Table 2-10. Two-Input AND Function Truth Table

Inputs		Output
A	B	Z
0	0	0
0	1	0
1	0	0
1	1	1

The following is an example of using **AND** logic in process control: If the liquid level in a process tank is high, and the inlet feed pump to the tank is running, design a logic circuit to open the tank outlet valve using electric relays. Figure 2-4 shows the relay ladder logic diagram for performing the required **AND** function in the application.

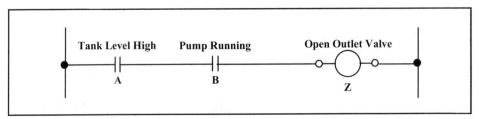

Figure 2-4. Relay-based AND logic control application.

The logic expressions for the **AND** function are as follows:

$$A \bullet B \bullet C = (A \bullet B) \bullet C = A \bullet (B \bullet C) \qquad (2\text{-}10)$$

$$A \bullet B = B \bullet A \qquad (2\text{-}11)$$

$$A \bullet A = A \qquad (2\text{-}12)$$

$$A \bullet 1 = A \qquad (2\text{-}13)$$

$$A \bullet 0 = 0 \qquad (2\text{-}14)$$

These identities can be verified by referring to the definitions of the **AND** gate and by using a truth table for the **AND** gate.

For example, Equation 2-13 ($A \bullet 1 = A$) can be verified as follows: let $B = 1$ and tabulate $A \bullet B = Z$ or $A \bullet 1 = Z$ in a truth table as follows.

Inputs	Output
A B	Z
0 1	0
1 1	1

Note that $A = Z$ in this truth table, so $A \bullet 1 = A$ for both values (0 or 1) of A.

Some important auxiliary identities used in logic design are as follows:

$$A + A \bullet B = A \qquad (2\text{-}15)$$

$$A + \overline{A} \bullet B = A + B \qquad (2\text{-}16)$$

$$(A + B) \bullet (A + C) = A + B \bullet C \qquad (2\text{-}17)$$

These identities are important because they help reduce the number of logic elements or gates required to implement a logic function.

NOR Function

Another common logic gate is the **NOR** gate, and Table 2-10 shows its operation for two inputs. It produces a logic 1 result if and only if all inputs are logic 0. Notice that the output of the **NOR** function is the opposite of the output of an **OR** gate.

Table 2-11. Two-Input NOR Function Truth Table

Inputs	Output
A B	Z
0 0	1
0 1	0
1 0	0
1 1	0

NAND Function

Another logic operation that is of interest is the **NAND** gate. Its operations
are summarized in Table 2-11 for a two-input **NAND** function. Note that
the output of the **NAND** function is the exact opposite of the **AND**
function output. When both inputs to the **NAND** are 1, the output is 0. In
all other configurations, the **NAND** function output is 1.

Table 2-12. Two-Input NAND Function Truth Table

Inputs	Output
A B	Z
0 0	1
0 1	1
1 0	1
1 1	0

Logic Function Symbols

Two common sets of symbols are used in process control applications to
represent logic function: *graphic* and *ladder*. The graphic symbols are
generally used on engineering drawings to convey the overall logic plan
for a discrete or batch control system. They are based on ANSI/ISA S5.2
Binary Logic Diagrams for Process Operations. The ladder logic symbols
are used to describe a logic control plan if you are using relays to
implement the control system or if you are using programmable controller
ladder logic to implement the control plan. Figure 2-5 compares these two
sets of symbols for the most common logic functions encountered in logic
control design.

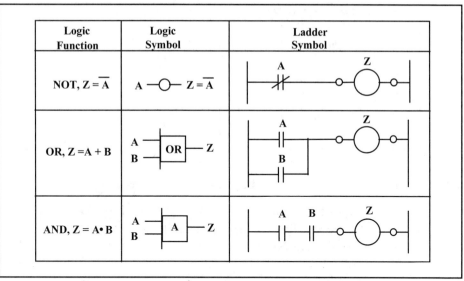

Figure 2-5. Comparison of logic symbols.

Ladder Logic Diagrams

Ladder diagrams are a traditional method for describing electrical logic controls. These circuits get their name from their resemblance to ladders with rungs. Each rung of a ladder is numbered, so we can easily cross-reference between the sections on the drawing that describe the control.

We can appreciate the utility of ladder logic diagrams by investigating a simple control example. Figure 2-6 shows a typical process level control application. In this application, let us assume that the flow into the tank is random. Let's also assume that we need to control the level in the tank by opening or closing the outlet valve (LV-1) based on the level sensed in the tank by a level switch (LSH-1). We will also provide the operator with a three-position hand, off, and automatic (HOA) switch to manually turn the valve ON or OFF or the option of selecting automatic control using the level switch to maintain the proper level in the tank.

The ladder logic design for this application is shown in Figure 2-7. If the panel-mounted HOA switch is in the automatic position and the level switch is closed, the solenoid will be energized. This is a simple example of a logical **AND** function in process control. If the HOA switch is in the hand or manual position, the valve will also be turned on.

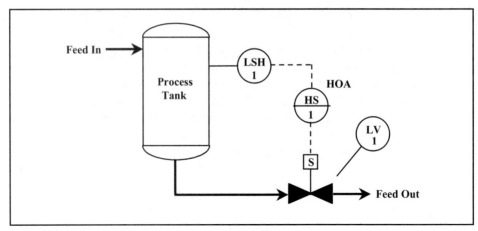

Figure 2-6. On/off control of process tank level.

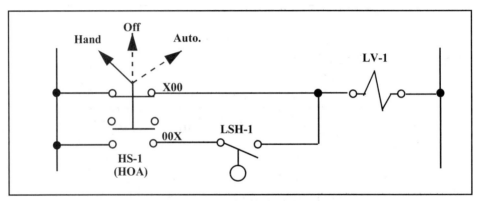

Figure 2-7. Ladder diagram for process tank-level control.

Suppose we designate the logic variables for the ladder diagram of Figure 2-7 as follows: A for hand position of HOA switch, B for automatic position of the HOA switch, C for the status of level switch LSH-1, and Z for the output to turn on solenoid valve LV-1. In that case, the logic equation for the control circuit is $Z = A + BC$. We can implement this control circuit with logic gates, as shown in Figure 2-8.

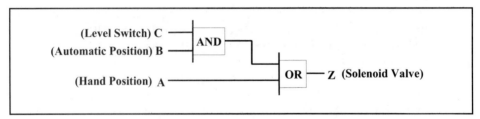

Figure 2-8. Logic gate implementation of tank-level control.

A more complex application might be to control tank liquid level between two level switches, a level switch high (LSH) and a level switch low (LSL). In this application, a pump supplying liquid to a tank is turned on and off to maintain the liquid level in the tank between the two level switches. The process is shown in Figure 2-9. It uses steam at a regulated flow to boil down a liquid to produce a more concentrated solution. This solution is then drained off periodically by opening a manual valve on the bottom of the tank. Note that a small diamond symbol is used to indicate that the pump is interlocked with the level switches on the tank.

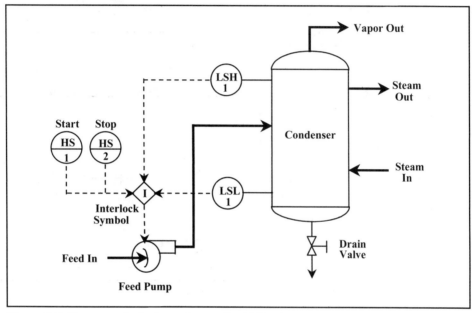

Figure 2-9. Pump control of condenser liquid level.

The electrical ladder diagram used to control the feed pump and hence the liquid level in the process tank is shown in Figure 2-10. To explain the logic of the control system, we will assume that the tank is empty and the low-level and high-level switches are closed. When a level switch is activated, the normally open contacts are closed and the normally closed contacts are opened. To start the control system, the operator depresses the start pushbutton (HS1). This energizes the control relay (CR1), which seals in the start pushbutton with the first set of contacts, which are denoted as CR1(1) on the ladder diagram. At the same time, the second control relay (CR2) is energized in rung 3 through contacts on LSH-1, LSL-1, and CR1. This turns on the pump motor starter MS1. The first set of contacts on CR2 (1) is used to seal in the low-level switch contacts, so that when the level in the tank rises above the position of the low-level switch on the tank, the pump will stay on until the liquid level reaches the high-level switch.

After the high-level switch is activated, the pump will be turned off. The system will now cycle on and off between the high and low levels until the operator depresses the stop pushbutton (HS2).

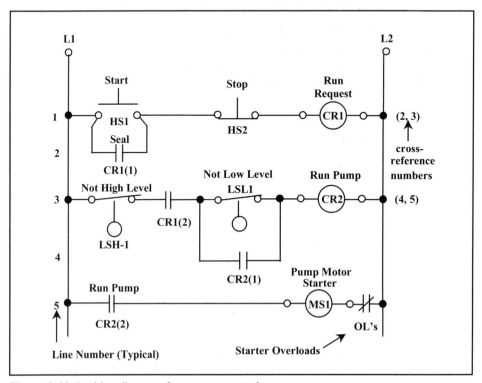

Figure 2-10. Ladder diagram for pump control.

This pump control application is a typical example of logic control in the process industries. You can implement the logic control system using a hardwired relay-based logic system, a programmable logic controller, or a distributed control system.

EXERCISES

2.1   Convert the binary number 1010 into a decimal number.

2.2   Convert the binary number 10011 into a decimal number.

2.3   Find the octal equivalent of the binary number 10101011.

2.4   Find the octal equivalent of the binary number 101111101000.

2.5   Convert the decimal number 33_{10} into its octal equivalent.

2.6   Convert the decimal number 451_{10} into its octal equivalent.

2.7   Convert the hexadecimal number 47_{16} into its decimal equivalents.

2.8   Convert the hexadecimal number 157_{16} into its decimal equivalents.

2.9   Convert the decimal number 56_{10} into its hexadecimal equivalent.

2.10   Convert the decimal number 37_{10} into its equivalent BCD code.

2.11   Convert the decimal number 270_{10} into its equivalent BCD code.

2.12   Express the words *Level Low* using ASCII code. Use hex notation.

2.13   Verify the logic identity $A + 1 = 1$ using a two-input **OR** truth table.

2.14   Write the logic equation for the start/stop control logic in lines 1 and 2 of the ladder diagram for the pump control application shown in Figure 2-10.

2.15   Write the logic equation for the control logic in lines 3 and 4 of the ladder diagram for the pump control shown in Figure 2-10.

2.16   Draw a logic gate circuit for the standard start/stop circuit used in the pump control application shown in Figure 2-10.

BIBLIOGRAPHY

1.   Boylestad, R. L., and L. Nashelsky. *Electronic Devices and Circuit Theory*, 3d ed. (Prentice-Hall, 1982).

2.   Budak, A. *Passive and Active Network Analysis and Synthesis* (Houghton Mifflin, 1974).

3.   Clare, C. R. *Designing Logic Systems Using State Machines* (McGraw-Hill, 1973).

4.   Floyd, T. L. *Digital Logic Fundamentals* (Charles E. Merrill, 1977).

5.   Grob, B. *Basic Electronics*, 5th ed. (McGraw-Hill, 1984).

6.   Kintner, P. M. *Electronic Digital Techniques* (McGraw-Hill, 1968).

7.   Malvino, A. P. *Digital Computer Electronics—An Introduction to Microcomputers*, 2d ed. (McGraw-Hill, 1983).

3

Electrical and Electronic Fundamentals

Introduction

The design and maintenance of programmable controller systems requires knowledge of the basic principles of electricity and electronics. This chapter will discuss the fundamentals of electricity and then investigate the electrical and electronic circuits that are commonly encountered in the instrumentation and control field. We will also discuss the operation and purpose of electrical and electronic devices, such as power supplies, relays, solenoid valves, contactors, control switches, and indicating lights.

Fundamentals of Electricity

Electricity is a fundamental force in nature that can produce heat, motion, light, and many other physical effects. This force is an attraction or repulsion existing between electric charges called *electrons* and *protons*. An electron is a small atomic particle that has a negative electric charge. A proton is a basic atomic particle with a positive charge. It is the arrangement of electrons and protons that determines the electrical characteristics of all substances. As an example, this page of paper has electrons and protons in it, but there is no evidence of electricity because the number of electrons equals the number of protons. In this case, the opposite electrical charges cancel, making the paper electrically neutral.

If we want to use the electrical forces associated with positive and negative charges, work must be done to separate the electrons and protons. For example, an electric battery can do such work because its chemical energy separates electrical charges, producing an excess of electrons at its negative terminal and an excess of protons at its positive terminal.

The basic terms encountered in electricity are electric charge (Q), current (I), voltage (v), and resistance (R). For common applications of electricity a charge of a very large number of electrons and protons is required. Therefore, it was convenient to define a practical unit called the coulomb (C) as being equal to the charge of 6.25×10^{18} electrons or protons. As mentioned, the symbol for electric charge is Q, which stands for quantity of charge. A charge of 6.25×10^{18} electrons or protons is stated as $Q = 1\ C$. Voltage or electrical potential difference is defined as the work required to move an charge in an electric field. When one joule of work is required to move one coulomb between two points in the electric field, the potential difference is one volt (v).

When the potential difference between two different charges causes a third charge to move, the charge in motion is called *electric current*. If the charges move at the rate of one coulomb per second past a given point, the amount of current is defined as one ampere (a). In equation form, $I = dQ/dt$, where I is the instantaneous current in amperes, and dQ is the differential amount of charge in coulombs passing a given point during the time period (dt) in seconds. If the current flow is constant, it is simply given by $I = Q/t$, where Q is the amount of current flowing past a given point in t seconds. Let's consider an example.

EXAMPLE 3-1

Problem: A steady flow of 12 coulombs of charge passes a given point in a copper conductor every three seconds. What is the current flow?

Solution: Since constant current flow is defined as the amount of charge (Q) that passes a given point per period of time (t) in seconds, we have

$$I = Q/t = 12C/3s$$

$$I = 4\ amps$$

The fact that a conductor carrying electric current can become hot is evidence that the work done by the applied voltage in producing current must be meeting some form of opposition. This opposition, which limits current flow, is called *resistance*.

Conductivity, Resistivity, and Ohm's Law

Some material has an important physical property called *conductivity*, that is, the ability to pass electric current. Suppose we have an electric wire (conductor) of length L and cross-sectional area A e apply a voltage V

between the ends of the wire. If V is in volts and L is in meters, we can define the voltage gradient (E), as follows:

$$E = \frac{V}{L} \text{ (volt/meter)} \tag{3-1}$$

Now, if a current I in amperes flows through a wire of area A in meters squared (m^2), we can define the current density J as follows:

$$J = \frac{I}{A} \text{ (amps/meters}^2) \tag{3-2}$$

The conductivity C is defined as the current density per unit voltage gradient E. Or in equation form we have

$$C = \frac{J}{E} \text{ (amps/meter}^2) / \text{(volts/meter)} \tag{3-3}$$

or

$$C = \frac{I/A}{V/L}$$

Resistivity (r) is defined as the inverse of conductivity, or

$$r = \frac{1}{C} \tag{3-4}$$

The fact that resistivity is a natural property of certain materials leads to the basic principle of electricity called Ohm's law.

Consider a wire of length L and area A. If it has resistivity, r, then its resistance R is

$$R = r\frac{L}{A} \tag{3-5}$$

The unit of resistance is the ohm t is denoted by the Greek letter omega, Ω. Since resistivity, r, is the reciprocal of conductivity, we obtain the following:

$$r = \frac{V/L}{I/A} \tag{3-6}$$

When Equation 3-6 is substituted into Equation 3-5, we obtain

$$R = \frac{V}{I} \qquad\qquad (3\text{-}7)$$

This relationship, $R = V/I$ or $I = V/R$ or $V = IR$, is called Ohm's law. It assumes that the resistance of the material used to carry the current flow will have a linear relationship between the voltage applied and the current flow. That is, if the voltage across the resistance is doubled, the current through it also doubles. The resistance of materials like carbon, aluminum, copper, silver, gold, and iron is linear and follows Ohm's law. Carbon is the material most commonly used to manufacture a device with fixed resistance. Such a device is called a *resistor*.

Wire Resistance

To compare the resistance and size of one conductor with another, a standard or unit size of conductor was established. A convenient unit of measurement for the diameter of a circular wire is the mil (0.001 inch) and a convenient unit of wire length is the foot. The standard unit of wire size in most cases is the *mil-foot*; that is, a wire is said to have a unit size if it has a diameter of 1 mil and a length of 1 foot.

In U.S. wire-sizing tables, the *circular mil* is the standard unit for the cross-sectional area of wire. Because the diameter of circular wire is normally only a small fraction of an inch, it is convenient to express these diameters in mils to avoid using decimals. For example, the diameter of a 0.010-inch conductor is expressed as 10 mils instead of 0.010 inch. The circular mil area is defined as the square of the mil diameter of a conductor, or area (cmil) $= d^2(\text{mil}^2)$. Example 3-2 shows how to calculate conductor area in circular mils.

EXAMPLE 3-2

Problem: Calculate the area in circular mils of a conductor with a diameter of 0.002 inch.

Solution: First, we must convert the diameter in inches into a diameter in mils. Since we defined 0.001 in. = 1 mil, this implies that 0.002 in. = 2 mils, so the circular mil area is $(2 \text{ mil})^2$ or 4 cmil.

A *circular-mil per foot* is a unit conductor that is one foot in length and has a cross-sectional area of one circular mil. The cmil per foot is useful when you are comparing the resistivity of various conductors. The specific resistance (*r*) in cmil-ohms per foot of some common solid metals at 20°C is given in Table 3-1.

Table 3-1. Specific Resistance (*r*) at 20°C

Material	r, cmil-Ω/ft
Silver	9.8
Copper (drawn)	10.37
Gold	14.70
Aluminum	17.02
Tungsten	33.20
Brass	42.10
Steel	95.80

Conductors that are used to carry electric current are normally manufactured out of copper because of its low resistance and relatively low cost. We can use the specific resistance (*r*) given in Table 3-1 to find the resistance of a conductor of length (*L*) in feet and of cross-sectional area (*A*) in cmil by using Equation 3-5, *R=rL/A*. Let's consider another example.

EXAMPLE 3-3

Problem: Find the resistance of 1,000 feet of copper (drawn) wire that has a cross-sectional area of 10,370 cmil and a wire temperature of 20°C.

Solution: From Table 3-1, the specific resistance of copper wire is 10.37 cmil-ohms/ft. Substituting the known values in the equation, the resistance is determined as follows:

$$R = r\frac{L}{A}$$

R = (10.37 cmil - Ω0/ft)(1000 ft)/(10370 cmil)

R = 1Ω.

You can use the equation for the resistance of a conductor, $R = rL/A$, in many applications. For example, if *R*, *r*, and *A* are known, you can determine the length by a simple mathematical transformation of the equation for the resistance of a conductor. A typical application of this is locating a problem ground point in a telephone line. The telephone company uses specially designed test equipment to find ground faults.

This equipment operates on the principle that the resistance of a conductor varies directly with length such that the distance between a test point and a fault can be measured accurately with properly designed equipment. Consider Example 3-4.

EXAMPLE 3-4

Problem: The resistance to ground on a faulty underground telephone line is 5 Ω. Calculate the distance to the point where the wire is shorted to ground. Assume that the line is a copper conductor with a cross-sectional area of 1,020 cmil and that the ambient temperature of the conductor is 20°C.

Solution: To calculate the distance to the point where the wire is shorted to ground, we use $L = RA/r$. Since $R = 5\Omega$, $A = 1,020$ cmils, and $r = 10.37$ cmil-Ω/ft, we have

$L = RA/r$

$L = (5\Omega) (1020 \text{ cmils})/10.37 \text{ cmil} - \Omega/\text{ft}$

$L = 492 \text{ feet}$

Wire Gauge Sizes

Electrical conductors are manufactured in sizes that are numbered according to a system known as the American Wire Gauge (AWG). As Table 3-2 shows, the wire diameters become smaller as the gauge numbers increase, and the resistance per 1,000 feet increases as the wire diameter decreases. The largest wire size listed in the table is 4, and the smallest is 24. This is the normal range of wire sizes typically encountered in process control applications. The complete AWG table goes from 0000 to 40. The larger and smaller sizes not listed in Table 3-2 are manufactured but are not commonly encountered in process control. Consider Example 3-5.

Table 3-2. American Wire Gauge for Copper Wire

AWG Number	Diameter, mil	Area cmil	Ω/1000 ft at 25°C	Ω/1000 ft at 65°C
08	128.0	16500	0.641	0.739
10	102.0	10400	1.020	1.180
12	81.0	6530	1.620	1.870
14	64.0	4110	2.58	2.97
16	51.0	2580	4.09	4.73
18	40.0	1620	6.51	7.51
20	32.0	1020	10.4	11.9
22	25.3	642	16.5	19.0
24	20.1	404	26.2	30.2

EXAMPLE 3-5

Problem: Determine the resistance of 2,500 feet of 14 AWG copper wire. Assume that the wire temperature is 25°C.

Solution: Using Table 3-2, we see that 14 AWG wire has a resistance of 2.58 Ω per 1,000 ft at 25°C. So the resistance of 2,500 ft is calculated as follows:

$$R = (2.58\Omega/1000 \text{ ft})(2500 \text{ ft}) = 6.5\Omega$$

Direct and Alternating Current

You will encounter two basic types of voltage signals in process control and measurement: direct current (dc) and alternating current (ac). In direct current, the flow of charges is in just one direction. A battery is one example of a dc power source. A graph of a dc voltage versus time is shown in Figure 3-1a, and the symbol for a battery is shown in Figure 3-1b.

An alternating voltage source periodically reverses its polarity. Therefore, the resulting current flow in a closed circuit will reverse direction periodically. Figure 3-2a shows a sine-wave example of an ac voltage signal. The schematic symbol for an ac power source is shown in Figure 3-2b.

The 60-cycle ac power used in homes and industry in the United States is a common example of ac power. The term *60 cycle* means that the voltage polarity and current direction go through 60 reversals or changes per second. The signal is said to have a frequency of 60 cycles per second. The unit for one cycle per second (cps) is 1 hertz (Hz). Therefore, a 60-cycle-per-second signal has a frequency of 60 Hz.

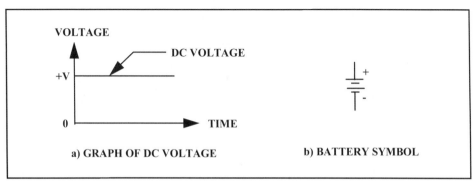

Figure 3-1. Steady dc voltage.

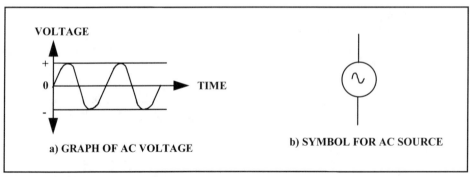

Figure 3-2. Sine-wave ac voltage.

Series Resistance Circuits

The components in a circuit form a *series circuit* when they are connected in successive order with the end of a component joined to the end of the next element. An example of a series circuit is shown in Figure 3-3.

In this circuit, the current (I) flows from the negative terminal of the battery through the two resistors (R_1 and R_2) and back to the positive terminal. According to Ohm's law, the amount of current (I) flowing between two points in a circuit equals the potential difference (V) divided by the resistance (R) between these points. V_1 is the voltage drop across R_1, V_2 is defined as the voltage drop across R_2, and the current (I) flows through both R_1 and R_2. From Ohm's law we therefore obtain the following:

$$V_1 = IR_1 \text{ and } V_2 = IR_2$$

so that

$$V_t = IR_1 + IR_2$$

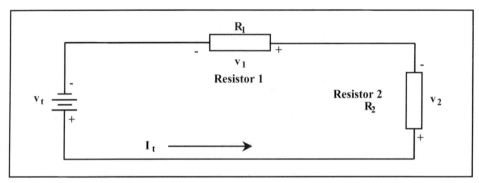

Figure 3-3. Series resistance circuit.

If we divided both sides of this equation by the current, I, in the series circuit, we obtain the following:

$$V_t/I = R_1 + R_2$$

The total resistance of the series circuit is defined by Ohm's law as the total voltage applied divided by the current in the circuit, or $R_t = V_t/I$. Therefore, the total resistance of the circuit is given by

$$R_t = R_1 + R_2$$

We can derive a more general equation for any number of resistors in series by using the classical *Law of Conservation of Energy*. According to this law, the energy or power supplied to a series circuit must equal the power dissipated in the resistors in the circuit. Thus,

$$P_t = P_1 + P_2 + P_3 \ldots P_n$$

Since power in a resistive circuit is given by $P = I^2R$, we obtain

$$I^2R_t = I^2R_1 + I^2R_2 + I^2R_3 + \ldots + I^2R_n$$

If we divide this equation by I^2, we obtain the series resistance formula for a circuit with n resistors:

$$R_t = R_1 + R_2 + R_3 + \ldots + R_n \tag{3-8}$$

EXAMPLE 3-6

Problem: Assume that the battery voltage in the series circuit of Figure 3-3 is 6 vdc and that resistor R_1 = 1K Ω and resistor R_2 = 2K Ω, where K is 1,000 or 10^3. Find the total current flow (I_t) in the circuit and the voltage across R_1 and R_2.

Solution: To calculate the circuit current I_t, first find the total circuit resistance R_t using Equation 3-7:

$$R_t = R_1 + R_2$$
$$R_t = 1k\Omega + 2k\Omega$$
$$R_t = 3k\Omega$$

Now, according to Ohm's law:

$$I_t = V_t/R_t$$
$$I_t = 6V/3K \ \Omega = 2 \text{ mA}$$

EXAMPLE 3-6 continued

The voltage across R_t (i.e., V_1) is given by

$V_1 = R_1 I_t$
$V_1 = (1K \ \Omega)(2 \ mA) = 2 \ v$

The voltage across R_2 (i.e., V_2) is obtained as follows:

$V_2 = R_2 I_t$
$V_2 = (2K \ \Omega) \ (2mA) = 4 \ v$

Parallel Resistance Circuits

When two or more components are connected across a power source, they form a parallel circuit. Each parallel path is called a *branch*, and each branch has its own current. In other words, parallel circuits have one common voltage across all the branches but individual currents in each branch. These characteristics are in contrast to series circuits, which have one common current but individual voltage drops.

A parallel resistance circuit with two resistors across a battery is shown in Figure 3-4. You can find the total resistance (R_t) across the power supply by dividing the voltage across the parallel resistance by the total current into the two branches. In the circuit of Figure 3-4, total current is given by $I_t = I_1 + I_2$. The current in branch 1 is given by $I_1 = V_t/R_1$, and the current in branch 2 is given by $I_2 = V_t/R_2$, so that $I_t = V_t/R_1 + V_t/R_2$. Since according to Ohm's law total current is given by $I_t = V_t/R_t$, we obtain

$$\frac{1}{R_t} = \frac{1}{R_1} + \frac{1}{R_2} \qquad (3\text{-}9)$$

or

$$R_t = \frac{R_1 \times R_2}{R_1 + R_2}$$

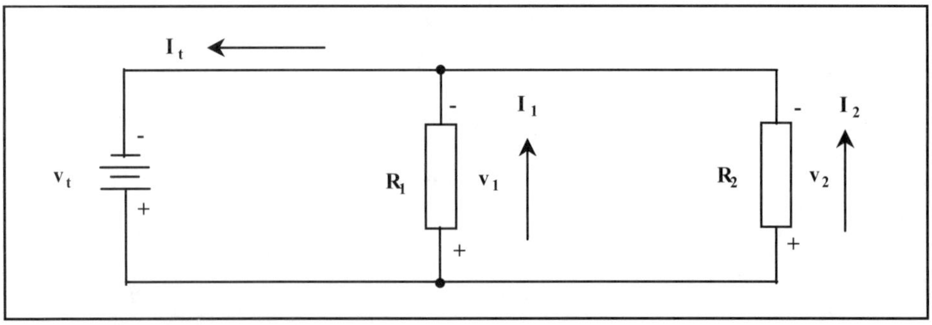

Figure 3-4. Parallel resistance circuit.

We can derive the general reciprocal resistance formula for any number of resistors in parallel from the fact that the total current I_t is the sum of all of the branch currents, or

$$I_t = I_1 + I_2 + I_3 + \ldots + I_n$$

Since current flow is defined as $I = dQ/dt$, this is simply the law of electrical charge conservation:

$$\frac{dQt}{dt} = \frac{dQ1}{dt} + \frac{dQ2}{dt} + \frac{dQ3}{dt} + \ldots + \frac{dQn}{dt}$$

This law implies that, since no charge accumulates at any point in the circuit, the differential charge, dQ, from the power source in the differential time period, dt, must appear as charges $dQ_1, dQ_2, dQ_3 \ldots dQ_n$ through the resistors $R_1, R_2, R_3 \ldots R_n$ at the same time. Since the voltage across each branch is the applied voltage V_t and $I = V/R$, we obtain

$$\frac{V_t}{R_t} = \frac{V_t}{R_1} + \frac{V_t}{R_2} + \frac{V_t}{R_3} + \ldots + \frac{V_t}{R_n}$$

If we divide both sides of this equation by V_t, we obtain the total resistance of n resistors in parallel, as follows:

$$\frac{1}{R_t} = \frac{1}{R_1} + \frac{1}{R_2} + \frac{1}{R_3} + \ldots + \frac{1}{R_n} \qquad (3\text{-}10)$$

To illustrate the basic concepts of a parallel circuit, let's look at a sample problem.

EXAMPLE 3-7

Problem: Assume that the voltage for the parallel circuit shown in Figure 3-4 is 12 v and that we need to find the currents I_t, I_1, and I_2 and the parallel resistance R_t, given that $R_1 = 30K\ \Omega$ and $R_2 = 30K\ \Omega$.

Solution: We can find the parallel circuit resistance R_t by using Equation 3-9:

$$R_t = \frac{R_1 \times R_2}{R_1 + R_2} \qquad\qquad R_t = \frac{(30k)(30k)}{30k + 30k}\Omega = 15k\Omega$$

EXAMPLE 3-7 continued

The total current flow is given by the following:

$$I_t = \frac{V_t}{R_t} = \frac{12V}{15k\Omega} = 0.8mA$$

The current branch 1 (I_1) is obtained as follows:

$$I_1 = \frac{V_t}{R_1} = \frac{12V}{30k\Omega} = 0.4mA$$

Since $I_t = I_1 + I_2$, the current flow in branch 2 is given by

$$I_2 = I_t - I_1 = 0.8 \text{ mA} - 0.4 \text{ mA} = 0.4 \text{ mA}$$

Wheatstone Bridge Circuit

The Wheatstone bridge circuit, named for English physicist and inventor Sir Charles Wheatstone (1802-1875), was one of the first electrical measuring instruments that could accurately measure resistance. A typical Wheatstone bridge circuit is shown in Figure 3-5. The circuit has two parallel resistance branches, with two series resistors in each branch and a galvanometer (G) connected across the branches. The galvanometer is a very sensitive electric charge-measuring instrument. It is momentarily connected across the bridge between points a and b. If these points are at the same potential, the meter will not deflect. The purpose of the circuit is to have the voltage drops be balanced across the two parallel branches so

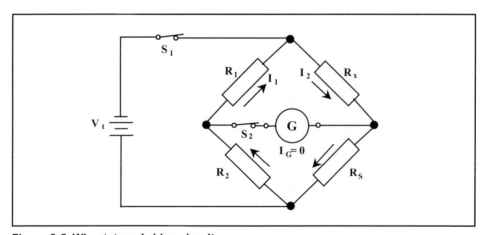

Figure 3-5. Wheatstone bridge circuit.

as to obtain zero volts across the meter. In the Wheatstone bridge, the unknown resistance R_x is balanced against a standard accurate resistor R_s to yield a precise measurement of resistance.

In the circuit shown in Figure 3-5, the switch S_1 is closed so the voltage V_t is applied to the four resistors in the bridge. To balance the circuit, the value of R_x is varied until a zero reading is obtained on the meter that has switch S_2 closed.

A typical application of the Wheatstone bridge circuit is to replace the unknown resistor R_x with a resistance temperature device (RTD). An RTD is a device that varies its resistance with a change in temperature so the balancing resistor dial can be calibrated to read out a process temperature.

When the bridge circuit is balanced (i.e., no current through galvanometer G), the circuit can be analyzed as two series resistance strings in parallel. The equal voltage ratios in the two branches of the Wheatstone bridge can be stated as follows:

$$\frac{I_2 R_x}{I_2 R_s} = \frac{I_1 R_1}{I_1 R_2}$$

Note that I_1 and I_2 can be canceled out in the previous equation. When they are, we can invert R_s to the right side of the equation to find R_x as follows:

$$R_x = R_s \frac{R_1}{R_2} \qquad (3\text{-}11)$$

To illustrate how a Wheatstone bridge is used, let's consider the following example.

EXAMPLE 3-8

Problem: Assume a Wheatstone bridge with R_1 = 1K Ω and R_2 = 10K Ω. If R_s is adjusted to read 50 Ω when the bridge is balanced (i.e., galvanometer reading at zero), calculate the value of R_x.

Solution: We can use Equation 3-11 to determine R_x:

$$R_x = R_s \frac{R_1}{R_2}$$

$$R_x = 50\Omega \frac{1k\Omega}{10k\Omega} = 5\Omega$$

Instrumentation Current Loop

Our brief discussion of resistive circuits earlier in the chapter was leading to the discussion of instrumentation current loops provided in this section. The dc current loop shown in Figure 3-6 is used extensively in the instrumentation field to transmit process variables to indicators and controllers. It is also used to send control signals to field devices in order to manipulate process variables such as temperature, level, and flow. The standard current range used is 4 to 20 mA. This value is normally converted into 1 to 5 vdc by a 250 Ω resistor at the input to controllers and indicators that are normally high-input impedance ($Z_{in} > 10$ MΩ) electronic amplifiers that draw virtually no current.

There are two main advantages to using the 4- to 20-mA current loop. First, only two wires are required for each remotely mounted field transmitter, so a cost savings is realized on both labor and wire when installing field devices. The second advantage is that the current loop is not affected by electrical noise or changes in lead wire resistance caused by temperature changes.

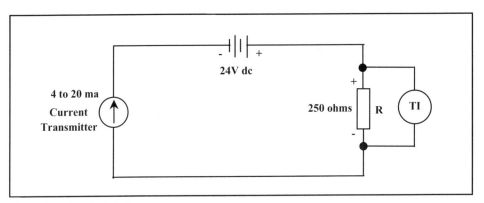

Figure 3-6. Typical 4- to 20-mA current loop.

Selecting Wire Size

Several factors must be considered when you are selecting the wire size in a programmable controller application. One factor is the permissible power loss ($P = I^2R$) in the electrical line. This power loss is electrical energy being converted into heat. If the heat produced is excessive, the conductors or system components might be damaged. Using large-diameter (low-wire-gauge) conductors will reduce the circuit resistance and, therefore, the power loss. However, larger-diameter conductors are more expensive than smaller ones and more costly to install. As a result, you will need to make design calculations to select the proper wire size.

A second factor to consider when selecting wire size is the resistance of the wires in the circuit. For example, let's assume we are sending a full-range 20-mA dc signal from a field instrument to a programmable controller input module located 2,000 feet away, as shown in Figure 3-7. Assuming that we are using 20 AWG wire, we can easily calculate the wire resistance (R_w). Using data from Table 3-2, we see that 20 AWG wire has a resistance of 10.4 Ω per 1,000 feet at 25°C. So the resistance for the 4,000 feet of wire is calculated as follows:

$$R_w = (10.4\Omega/1000 \text{ ft})(4000 \text{ ft}) = 41.6\Omega$$

This resistance increases the load on the 24 vdc power supply that is used to drive the current loop.

Using Ohm's law you can calculate the voltage drop V_w in the wire when the field instrument is sending 20 mA of current to the programmable controller analog input module, as follows:

$$V_w = IR_w = (20 \text{ mA})(41.6\Omega) = 0.832 \text{ volt}$$

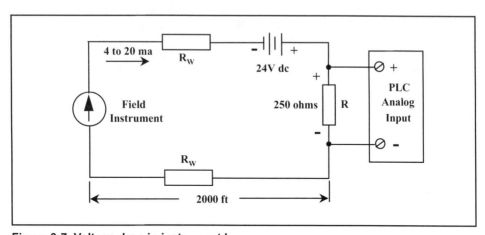

Figure 3-7. Voltage drop in instrument loop.

A third factor to consider when selecting wire size is the current-carrying ability of the conductor. When current flows in a wire, heat is generated. The temperature of the wire will rise until the heat that is radiated away, or otherwise dissipated, is equal to the heat generated in the conductor. If the conductor is insulated, it is not as easy to remove the heat produced as it would be if the conductor were not insulated. So, to protect the insulation from excessive heat, you must keep the current flowing in the wire below a certain value. Rubber insulation will start to deteriorate at relatively low temperatures. Teflon and certain plastic insulations retain their insulating properties at higher temperatures, and insulations such as asbestos are effective at still higher temperatures.

Electrical cables might be installed in locations where the ambient temperature is relatively high. In these cases, the heat produced by external sources adds to the total heating of the electrical conductor. Therefore, you must make allowances for the ambient heat sources encountered in industrial environments when designing a programmable controller system. The maximum allowable operating temperature of insulated wires and cables is specified in the manufacturer's electrical design tables.

Table 3-3 gives an example of the maximum allowable current-carrying capacities of copper conductor with three different types of insulation. The current ratings listed in the table are those permitted by the National Electrical Code. You can use this table to determine the safe and proper wire size in an electrical wiring application. Example 3-9 will help to illustrate a typical wire-sizing calculation.

Table 3-3. Ampacities of Insulated Copper Conductor National Electrical Code

Conductor Size (AWG)	60°C (140°F) Types: TW, UF	75°C (167°F) Types: FEPW, RH, RHW, THHW, THW, THWN, XHHW, USE, ZW	85°C (185°F) Type: V18
14	20	20	25
12	25	25	30
10	30	35	40
8	40	50	55
6	55	65	70

EXAMPLE 3-9

Problem: Calculate the proper wire size for the 120 vac feed to the programmable controller system shown in Figure 3-8, assuming a maximum ambient temperature of 60°C.

Solution: The total ac feed current is the sum of all the branch currents, or

$$I_t = I_1 + I_2 + I_3 + I_4$$

$$I_t = 2\,A + 8\,A + 5\,A + 5\,A$$

$$I_t = 20\ amps$$

Using Table 3-3, we can see that the main power conductors (hot, neutral, and ground) must be a minimum size of 14 AWG.

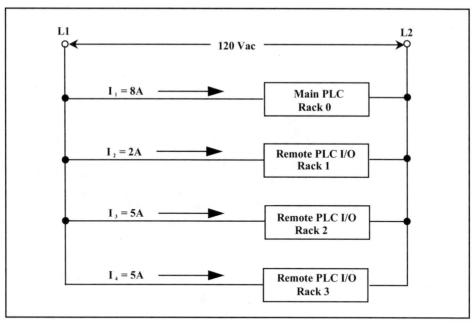

Figure 3-8. PLC wire-sizing application.

Power Supplies

It is important that you have a basic understanding of the design and operation of dc power supplies because they are used widely in control systems applications. DC power supplies use either half-wave or full-wave rectification depending on the power supply application. Figure 3-9 shows the output waveforms produced by half-wave and full-wave rectification. A schematic diagram of a half-wave rectifier is shown in Figure 3-10. In this circuit, the positive and negative cycles of the ac voltage across the secondary winding are in phase with the signal at the primary. This is indicated by the dots on each side of the transformer symbol.

Assume that the top of the transformer secondary is positive and the bottom is negative during a positive cycle of the input. In that case, the diode is forward-biased, and current flow is permitted through the load resistor, R_L, as indicated.

During the negative input cycle, the top of the transformer will be negative and the bottom will be positive. This voltage polarity reverse biases the diode. A reverse-biased diode represents an extremely high resistance and serves as an open circuit. Therefore, with no current flowing through R_L, the output voltage will be zero for this cycle.

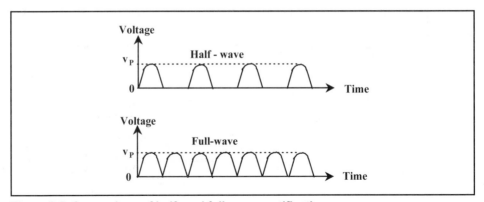

Figure 3-9. Comparison of half- and full-wave rectification.

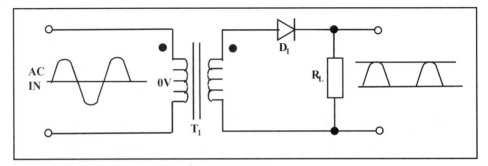

Figure 3-10. Half-wave rectifier circuit.

The basic full-wave rectifier using two diodes is shown in Figure 3-11. The cathodes of both diodes are connected to obtain a positive output. The anodes of each diode are connected to opposite ends of the transformer secondary winding. The load resistor (R_L) and the transformer center are connected to ground to complete the circuit.

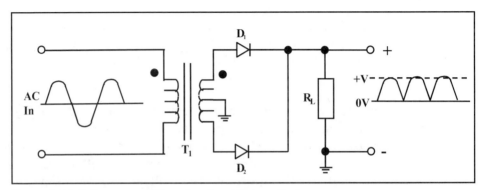

Figure 3-11. Full-wave rectifier circuit using two diodes.

In a full-wave rectifier, the current through R_L is in the same direction for each alteration of the input. As a result, we obtain dc output for both halves of the sine-wave input, or full-wave rectification.

The ripple frequency of the full-wave rectifier is also different from that of a half-wave circuit. Since each cycle of the input produces an output across R_L, the ripple frequency will be twice the input frequency. It is easier to filter the higher ripple frequency and characteristic output of the full-wave rectifier than it is a similar half-wave output. To reduce the amount of ripple, a filter capacitor is placed across the output. (A capacitor is two metal plates separated by a dielectric material.)

A bridge structure of four diodes is used in most power supplies to achieve full-wave rectification. In the bridge rectifier shown in Figure 3-12, two diodes will conduct during the positive alteration and two will conduct during the negative alteration. A bridge rectifier does not require the center-tapped transformer that is used in a two-diode, full-wave rectifier.

The dc output appearing across R_L of the bridge circuit has less ripple amplitude because a filter capacitor has been placed across the output. A capacitor opposes any changes in the voltage signal. If the filter capacitor is properly sized and placed across the output, it will prevent the output signal from returning to zero at the end of each half cycle, as shown in Figure 3-12.

Bridge rectifiers are commonly used in the electronic power supplies for instruments because of their simple operation and desirable output. The diode bridge is generally housed in a single enclosure that has two input and two output connections. The dc output voltage signal from a bridge rectifier is filtered to produce the smooth dc signal that most instrumentation circuits and devices need.

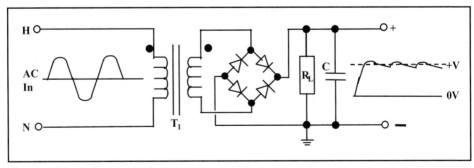

Figure 3-12. Full-wave bridge rectifier circuit.

We have now discussed the most common types of electronic devices and circuits encountered in programmable controller applications. Next, we need to discuss some common electrical control devices.

Electrical Control Devices

A wide variety of electrically operated devices are found in control applications. In this section we will cover the most common devices and give the electrical symbol used in drawings to represent each device.

Electrical Relays

The most common control device encountered in control applications is the *electromechanical relay*. It is called "electromechanical" because it consists of both electrically operated parts and mechanical components. One of its electrical parts is a long thin wire wound into a close-packed helix around an iron core. When a changing electric current is applied, a strong magnetic field is produced, and the resulting magnetic force moves the iron core, which is connected to a set of contacts. This transfers the common contact from the normally closed contact to the normally open contact.

A simplified representative diagram of an electrical relay with two sets of relay contacts is shown in Figure 3-13. Each set consists of a common (COM) contact, a normally open (NO) contact, and a normally closed (NC) contact. "Normally open" and "normally closed" refer to the status of contacts when no electrical power is applied to the relay or when the relay is in the shelf position. For example, if there is no electrical continuity between a common contact and another contact in the set when the relay is on a storage shelf, this set of contacts is termed the "normally open set of contacts." These contacts in the relay are used to make or break electrical connections in control circuits.

The electrical schematic symbol for an NO set of contacts is given in Figure 3-14a, and the symbol for the NC set is shown in Figure 3-14b. The standard symbol for the relay solenoid is a circle with two connection dots, as shown in Figure 3-14c. This symbol will normally also include a combination of letter(s) and number(s) to identify each control relay on a control drawing. Typical designations are *CR1, CR2, . . . CRn* for the control relays on a schematic.

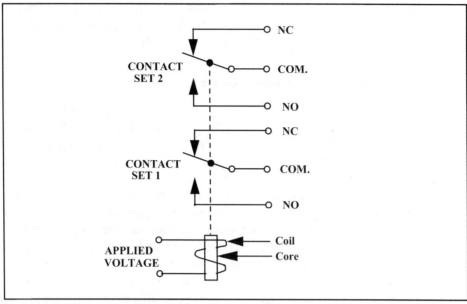

Figure 3-13. Pictorial representation of an electric relay.

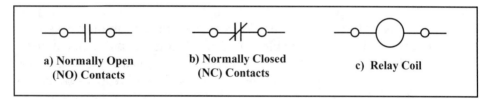

a) Normally Open
(NO) Contacts

b) Normally Closed
(NC) Contacts

c) Relay Coil

Figure 3-14. Electrical relay symbols.

Electric Solenoid Valves

Another electrically operated device commonly encountered in process control is the *solenoid valve*. It is a combination of two basic functional units: an electromagnetic solenoid with its core and a valve body containing one or more orifices. Flow through an orifice is allowed or prevented by the action of the core when the solenoid is energized or deenergized.

Solenoid valves normally have a solenoid mounted directly on the valve body. The core is enclosed and free to move in a sealed tube called a *core tube,* thus providing a compact assembly. The two different electrical schematic symbols used to represent solenoid valves on electrical and control drawings are given in Figure 3-15.

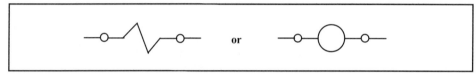

Figure 3-15. Schematic symbols used to represent solenoid valve.

The three common types of solenoid valves are *direct acting, internal pilot-operated,* and *manual reset.* In direct-acting valves, the solenoid core directly opens or closes the orifice, depending upon whether the solenoid is energized or deenergized. The valve will operate from 0 psi inlet pressure to its maximum rated inlet pressure. The force required to open a solenoid valve is proportional to the orifice size and the pressure drop across the valve. As the orifice size increases, the force needed to operate the valve increases. To keep the solenoid coil small and to open large orifices at the same time, internal pilot-operated solenoid valves are used.

Internal pilot-operated solenoid valves have a pilot and bleed orifice and rely on the line pressure to operate. When the solenoid is energized, the core opens the outlet side of the valve. The imbalance of pressure causes the line pressure to lift the piston or diaphragm off the main valve orifice, which opens the valve. When the solenoid is deenergized, the pilot orifice is closed. The full line pressure is applied to the top of the piston or diaphragm through the bleed orifice, closing the valve. In some applications, the bleed orifice is replaced by a small manual valve, which makes it possible to control how quickly the valve opens or closes.

The manual reset type of solenoid valve must be manually positioned (latched). It will return to its original position when the solenoid is energized or deenergized, depending on the valve design.

Three configurations of solenoid valves are normally available for process control applications: two-, three-, and four-way. Figure 3-16 shows the process diagram symbols used to represent two-, three-, and four-way solenoid valves. Two-way valves have one inlet and one outlet process connection. They are available in either normally open or normally closed configurations. Three-way solenoid valves have three process piping connections and two orifices (one orifice is always open, and the other is always closed). These valves are normally used in process control to alternately apply air pressure to and exhaust air from diaphragm-operated control valves or single-acting cylinders. Four-way solenoid valves have four pipe connections: one for supply air pressure, two for process control, and one for air exhaust. These valves are normally used to control double-acting actuators on control valves.

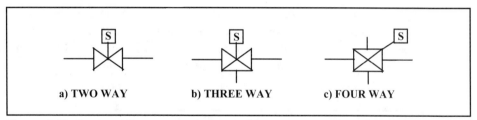

a) TWO WAY   **b) THREE WAY**   **c) FOUR WAY**

Figure 3-16. Process diagram symbols used to represent solenoid valves.

Electrical Contactors

Electrical contactors are relays that are able to switch high-current loads normally greater than 10 amps. They are used to repeatedly make and break electrical power circuits. An electrically operated solenoid is the most common operating mechanism for contactors. The solenoid is powered by low-ampere ac voltage signals from control circuits. Instead of opening or closing a valve, the linear action of the solenoid coil is used to open or close sets of contacts with high power ratings. The higher-rated sets of electrical contacts are used in turn to control loads, such as motors, pumps, and heaters. The schematic symbol for a typical electrical contactor is shown in Figure 3-17.

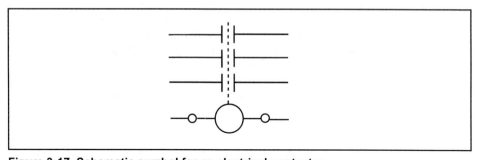

Figure 3-17. Schematic symbol for an electrical contactor.

Other Control Devices

Hand switches, instrument switches, control switches, pushbuttons, and indicating lights are a few of the other devices frequently encountered in process control. Switches and pushbuttons are used to control other devices, such as solenoid valves, electric motors, motor starters, heaters, and other process equipment.

Hand switches come in a variety of designs and styles, and they can be mounted on control panels or in the field near the process equipment. Instrumentation switches are generally field mounted and are used to sense pressure, temperature, flow, liquid level, and position.

To make it easier to read control drawings, the American National Standards Institute (ANSI) developed a set of standardized symbols. The most common symbols used in control diagrams are shown in Figure 3-18; they are based on ANSI Y32.2 for electrical controls.

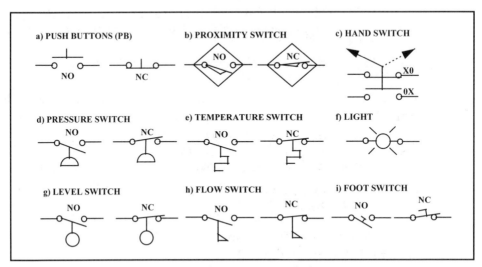

Figure 3-18. Common electrical and instrument symbols.

EXERCISES

3.1   A charge of 15 coulombs moves past a given point every second. What is the current flow?

3.2   Calculate the current in a conductor if 30×10^{18} electrons pass a given point in the wire every second.

3.3   Calculate the area in circular mils (cmil) of a conductor that has a diameter of 0.003 inch.

3.4   Calculate the resistance of 500 feet of copper wire with a cross-sectional area of 4,110 cmil if the ambient temperature is 20°C.

3.5   Calculate the resistance of 500 feet of silver wire with a diameter of 0.001 inch if the ambient temperature is 20°C.

3.6   The resistance to ground on a faulty underground telephone line is 20 Ω. Calculate the distance to the point where the wire is shorted to ground if the line is a 22 AWG copper conductor and the ambient temperature is 25°C.

3.7   Find the total current (I_t) in a circuit that has two 250 Ω resistors in series with a voltage source of 24 vdc. Also find the voltage drop across each resistor, where V_1 is voltage across the first resistor and V_2 is voltage across the other resistor.

3.8 In the parallel resistance circuit of Figure 3-4, assume that $R_1 = 100$ Ω, $R_2 = 200$ Ω, and $V_t = 100$ vdc. Find I_1, I_2, and I_t.

3.9 Assume that the Wheatstone bridge circuit shown in Figure 3-5 is balanced (i.e., $I_g = 0$) and that $R_1 = 2K$ Ω, $R_2 = 10K$ Ω, $R_s = 2K$ Ω, and $V_t = 12$ vdc. Calculate the value of I_1, I_2, and R_x. Also, what are the voltage drops across R_1 and R_2?

3.10 Explain the operation and purpose of electromechanical relays.

3.11 List the three common types of electrically operated solenoid valves.

BIBLIOGRAPHY

1. Bartkowiak, R. A. *Electric Circuit Analysis* (Harper & Row, 1985).

2. Bogart Jr., T. F. *Electric Circuits* (Macmillan, 1988).

3. Boylestad, R., and L. Nashelsky. *Electric Devices and Circuit Theory*, 3d ed. (Prentice-Hall, 1982).

4. Budak, A. *Passive and Active Network Analysis and Synthesis* (Houghton Mifflin, 1974).

5. Grob, B. *Basic Electronics*, 5th ed. (McGraw-Hill, 1984).

6. Hughes, T. A. *Measurement and Control Basics*, 2d ed. (ISA, 1995).

7. Malvino, A. P. *Electronic Principles*, 2d ed. (McGraw-Hill, 1979).

8. Motorola Semiconductor Products Inc. *Zener Diode Handbook* (Motorola Semiconductor Products, May 1967).

9. National Fire Protection Association. *The National Electrical Code Handbook*, 6th ed. (National Fire Protection Association, 1996).

Input/Output Systems

Introduction

The input/output (I/O) system provides the physical connection between the process equipment and the central processing unit (CPU), or simply the "processor." It uses interface circuits to convert the sensed or measured physical quantities of the process, such as motion, level, temperature, pressure, and position, into a logic signal to be used by the programmable logic controller (PLC) processor. Based on the status of the sensed or measured values, the control program in the PLC processor uses different output circuits or modules to activate devices, such as valves, motors, pumps, and alarms, to exercise control over a machine or process.

These I/O circuits or modules are mounted in equipment housings or, in the case of micro-PLCs, in part of the PLC housing. In most PLC housings, you can insert any I/O module into any I/O slot. The housings are designed so you can remove the I/O modules without turning off the ac power or removing the field wiring. Most I/O modules use printed circuit board technology, and the circuit boards have an edge connector that can be inserted into a plug in the backplane connector of the rack. This backplane is a printed circuit card that contains the parallel communications lines, or bus, to the processor and the dc voltages that are required to power the logic and interface circuits in the I/O modules.

Discrete Inputs

Discrete inputs are the most common class of field signals in a PLC system. A discrete-input field device provides an input signal that is separate and distinct in nature. These signals have only two states, OPEN/CLOSED or ON/OFF. Table 4-1 lists the discrete-input devices most often encountered in process control applications. Except for the photoelectric sensor, they all normally involve some form of switch contact or relay contact that is open or closed. The photoelectric sensor may have a relay contact output or an ON/OFF voltage signal such as 0 or 5 volts dc.

Table 4-1. Discrete Input Devices

Thumbwheel switches	Position switches
Temperature switches	Pressure switches
Flow switches	Hand switches
Level switches	Proximity switches
Valve position switches	Relay contacts
Starter auxiliary contacts	Limit switches
Selector switches	Motor starter contacts
Pushbuttons	Photoelectric sensors

If an input device is closed, the input circuit in the PLC senses the supplied voltage. To indicate the status of the device it then converts it into a logic-level signal that is acceptable to the CPU. A logic 1 indicates ON or CLOSED, and a logic 0 indicates OFF or OPENED.

Discrete Outputs

Discrete-output control is limited to devices that require that only one of two states be switched, such as ON/OFF, OPEN/CLOSED, or extended/retracted. Table 4-2 lists the most common discrete-output devices encountered in process and machine control applications.

Table 4-2. Discrete Output Devices

Annunciators	Electric valves
Alarm lights	Alarm horns
Electric control relays	Solenoid valves
Electric fans	Motor starters
Indicating lights	Heater starters

In operation, the PLC output interface circuit switches the control voltage that is connected to an output device. If an output is turned on through the control program, the interface circuit switches the control voltage to activate the referenced (addressed) output device.

I/O Signal Types

Each discrete input and output signal is powered by some field- or panel-supplied voltage source (e.g., +5 vdc, 120 vac, 24 vdc, etc.). I/O interface circuits are available for various ac and dc voltage ratings, as listed in Table 4-3.

The most common voltage levels used in industrial applications are 120 vac and 230 vac because they are readily available in industrial plants. However, low-voltage direct current sources, such as +5 vdc, +12 vdc, +24 vdc, and +48 vdc, are also widely used because they pose less risk of injury than do 120 or 230 vac.

Table 4-3. Typical Discrete I/O Signal Values

+ 5 volt dc	120 volt ac or dc
+ 12 volt dc	230 volt ac or dc
+ 24 volt ac or dc	100 volt dc
+ 48 volt ac or dc	Dry contacts

Sinking and Sourcing Operations

Sinking and sourcing operations refer to the electrical configuration of a device's electronic circuit, whether it is an input module or a field input device. If the device provides current during its ON state, the device is said to be sourcing current. Conversely, if the device receives current in the ON or TRUE state, it is said to be sinking current. Therefore, we can have sinking and sourcing field devices as well as sinking and sourcing input modules. However, the most common configuration using PLC applications is the sourcing field-input device and the sinking input module.

Potential interface problems can arise if you do not design the I/O system to properly match the sinking and sourcing operations of the devices in your system. You must use a sinking input module if the field devices connected to the module are sourcing devices. Conversely, for the system to operate properly, you must use sourcing input circuits if they are connected to a sinking field device. A problem would arise, for instance, if an input module were designed for sink operation and all input devices except one were operating in a source configuration. The sinking input device may be ON, but the module would not detect the ON signal even though a voltage could be measured across the module's input terminals. There is also the potential that the mismatched field device and the input circuit in the I/O module could be damaged.

Discrete AC Voltage Input Circuits

A block diagram of a typical alternating current (ac) voltage discrete-input circuit is shown in Figure 4-1. PLC manufacturers' discrete-input circuits vary widely, but, in general, discrete-input circuits operate much like the circuit shown in Figure 4-1.

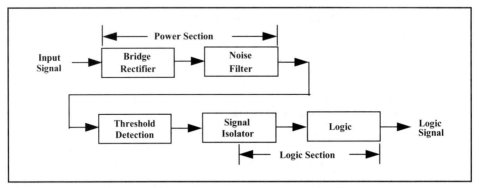

Figure 4-1. Block diagram of a typical discrete ac input circuit.

The input voltage circuit is composed of two primary parts: the power section and the logic section. The power and logic sections of the circuit are normally coupled with a circuit that electrically isolates the input power section from the logic circuits. This electrical isolation is very important in a normally noisy industrial environment. The main problem with the early application of computers in process control was that the input and output circuits were not designed for the harsh industrial environment.

Figure 4-2 shows a typical discrete ac input circuit. The power section of the circuit performs the function of converting the incoming voltage (115 vac, 230 vac, etc.) from a field-input device, as shown in Table 4-3, to a logic-level signal that the PLC processor can use. The bridge rectifier circuit converts the ac incoming signal into a dc level. The dc level is then sent to a filter circuit consisting of capacitor, C, and resistors, R_2 and R_3, which minimize the ripple from the full-wave bridge circuit. This RC filter circuit causes a signal delay of typically 10 to 25 milliseconds (msec). The threshold circuit uses a zener diode (Z_d) to detect whether the incoming signal has reached the proper voltage level for the specified input rating. If the input signal exceeds and remains above the threshold voltage for a duration that is at least as long as the filter delay, the signal will be accepted as a valid input.

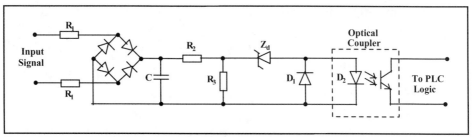

Figure 4-2. Typical ac voltage discrete-input circuit.

When a valid signal is detected, it is passed through to the isolation circuit, which completes the electrically isolated transition from ac or dc voltage to a logic-level voltage. The logic circuit uses the logic-level signal from the isolator, and the signal is made available to the processor via the data bus on the backplane of the PLC rack. Electrical isolation up to 1,500 vac is generally provided so there is no electrical connection between the field device (power) and the controller (logic). This electrical separation helps prevent large voltage spikes from damaging the logic side of the interface (or the controller). An optical coupler normally provides the coupling between the power and logic sections, as shown in Figure 4-2. Electrical isolation is one of the reasons the programmable controller has gained wide acceptance in the process industries.

In small, medium, and large PLC systems, these discrete-input circuits are mounted together on a single circuit board and installed in an input module. Input modules are available that have four, eight, sixteen, or thirty-two input circuits in one unit.

Discrete AC Input Modules

Most discrete ac input modules will have a signal indicator to signify that the proper input voltage level is present (a switch is closed). A light-emitting diode (LED) indicator is normally used to indicate the status of the input. These indicating lights are an important aid during system start-up and troubleshooting. A discrete ac input connection diagram is shown in Figure 4-3. The figure shows a 120-vac hot (L1) connected to the field devices, and a 120-vac neutral (L2) applied to the neutral (N) terminal of the input module. The term *ACI-120* above the module in Figure 4-3 is a typical model number that might be used by a PLC manufacturer. We will use similar model numbers for other I/O modules in this chapter.

Direct Current (DC) Input Modules

The dc voltage input modules convert discrete ON/OFF direct current inputs into logic-level signals that are compatible with the programmable controller. They are generally available in three voltage ranges: twelve, twenty-four, and forty-eight volts dc. Instruments that are typically compatible with the module include limit switches, valve position switches, pushbuttons, dc proximity switches, float switches, and photoelectric sensors.

The wiring diagram for a dc input module would be the same as for the ac input module shown in Figure 4-3 except the supply voltage would be a dc voltage instead of an ac voltage. The ac hot voltage signal (L1) sent to the field devices would be replaced by a positive dc voltage, and the neutral terminal on the module would be replaced by a dc common terminal.

Transistor-Transistor Logic (TTL) Input Modules

The TTL input modules allow the controller to accept signals from TTL-compatible devices, including solid-state controls and sensing instruments. TTL inputs are also used to interface with 5 vdc level control devices and several types of photoelectric sensors. The TTL interface is configured much like the dc input modules. However, the input delay time caused by filtering is generally much shorter. TTL input modules normally require an external +5 vdc power supply.

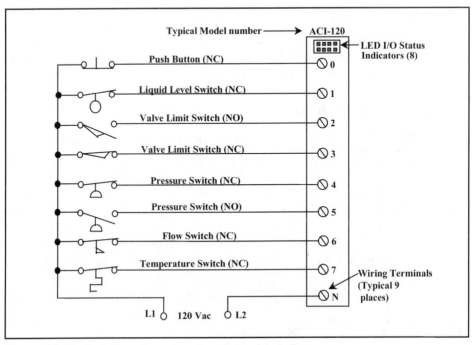

Figure 4-3. Typical discrete ac input module wiring diagram.

Isolated Discrete Input Module

Input and output modules usually have a common return line connection for each group of inputs or outputs on a single module. However, sometimes you may have to connect an input device that has different ground levels to the controller. In this case, isolated input modules with separate return lines for each input circuit are available (ac or dc) to accept these signals. The isolated interface and the standard discrete I/O operate the same except the common of each input is separated from the other commons in the module. The result is that the isolated input module requires twice as many input connection terminals. Consequently, the input module can accommodate only half the inputs in the same physical space (see Figure 4-4). The module number for the isolated 120-vac input module is ``IACI-120,'' and the isolated 220-vac input module model number is ``IACI-220.''

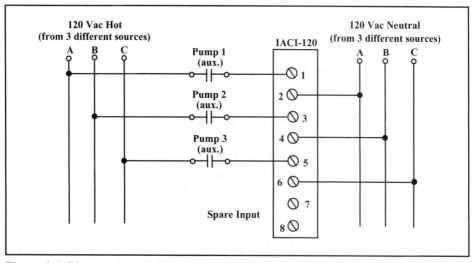

Figure 4-4. Discrete isolated ac input module wiring diagram.

Discrete AC Output Circuit

Figure 4-5 shows a block diagram for a typical discrete ac output circuit. AC output circuits vary widely among PLC manufacturers. The block diagram is representative of the basic operations performed in ac output circuits. The circuit consists primarily of the logic and power sections, coupled by an isolation circuit. The output interface can be thought of as a simple switch through which power can be provided to control the output device.

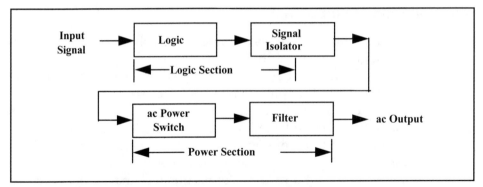

Figure 4-5. Block diagram of a discrete ac output circuit.

First, the processor sends an output signal of 0 or 1 to the logic circuit section. The signal from the logic section is then passed through an isolation circuit. The logic signal from the isolation circuit is next fed to an ac power-switching circuit and filter. Finally, this discrete ac output signal controls any ac-operated field device connected to the module output point.

The ac power-switching section generally uses a Triac or a silicon-controlled rectifier (SCR) to switch the ac power between ON or OFF. The ac switch is normally protected by an RC circuit or a metal oxide varistor (MOV), which is used to limit the peak voltage to some value below the maximum voltage rating. These protection circuits or devices also prevent electrical noise from affecting the module operation. A fuse may be provided in the output circuit to prevent excessive current from damaging the ac switch. If the fuse is not provided on each circuit in the module, it should be added to the exterior of each output circuit.

Discrete AC Output Module

In small, medium, and large PLC systems, discrete output ac circuits are mounted together on a single circuit board and installed in an output module. Output modules are available with four, eight, sixteen, or thirty-two output circuits on the circuit card.

As with input modules, the output modules also have light-emitting diode (LED) indicators to show the state of the operating logic. An ac output module connection diagram is illustrated in Figure 4-6. The LED indicators are located on the top of the module. The first two outputs on the module are wired to two starters for Heater 1 and Heater 2. The starter overload (OL) contacts are wired in series to turn off the starter in case there is a high current in the starter circuits. The next two outputs at terminals 2 and 3 are connected to 120-vac solenoid valves LV-1 and LV-2. The last four output points are connected by four starters with OLs in

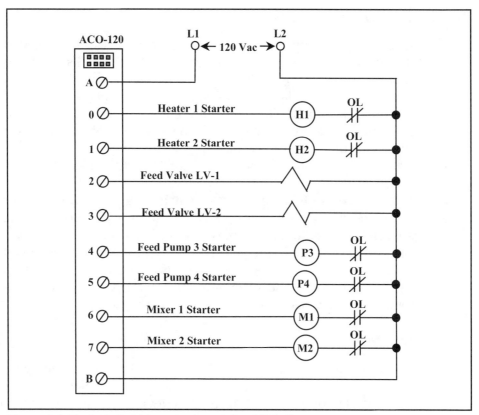

Figure 4-6. Typical discrete output module wiring diagram.

series with the starter coils. Note that the switching voltage is field supplied to the module so the module is a sourcing device.

Direct Current (DC) Output Module

The dc output module is used to switch direct current loads. The functional operation of the dc output is similar to that of the ac output. However, the power circuit generally employs a power transistor to switch the load. Like Triacs, transistors are also susceptible to excessive applied voltages and large surge currents, which could result in overheating and a short circuit. To prevent this from occurring, you would normally protect the power transistor with a fuse.

The wiring diagrams for a dc output module and the ac output module shown in Figure 4-6 would be the same, except the supply voltage would be a dc voltage instead of an ac voltage. The terminal of the ac hot line would be replaced by a positive dc voltage terminal, and the ac neutral terminal would be replaced by a ground or negative dc voltage terminal.

Dry Contact Output Module

The dry contact output module allows output devices to be turned ON or OFF by a normally open (NO) or normally closed (NC) set of relay contacts. The advantage of relay or dry contact outputs is that they provide electrical isolation between the PLC and the field device. The solid-state electrical switching circuit in the standard ac output module has a small leakage current even when the switching circuit is turned off. This small current can cause false signals in some cases. In these applications, you should use the dry contact output module.

Dry contact outputs can be used to switch either ac or dc loads. However, they are normally used in ac applications to provide electrical isolation between PLCs and other complex electrical equipment, such as variable speed drives (VSDs). Figure 4-7 shows a typical dry contact output module with four normally open contacts controlling the starting and stopping of two variable-speed motor drive units. In this application, there is complete electrical isolation between the PLC and VSDs.

TTL Output Module

The TTL output module allows the controller to drive output devices that are TTL-compatible, such as seven-segment LED displays, integrated circuits, and various +5 vdc logic-based devices. These modules generally require an external +5 vdc power supply with specific current requirements.

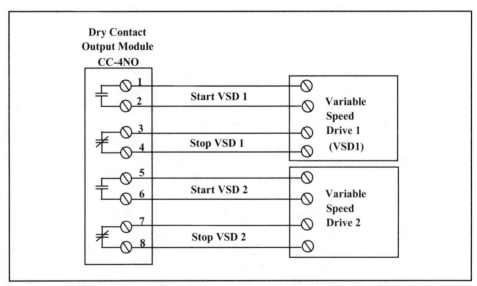

Figure 4-7. Typical dry contact output module wiring diagram.

Isolated AC Output Module

An isolated ac output interface is shown in Figure 4-8. Note that the isolated ac output module is driving three different loads (starters for Pumps 1, 2, and 3), which are connected to three different sources of ac power. The advantage of this module is that you do not have to be concerned about having different ac voltage sources in your process plant. The disadvantages are that they increase the amount of wiring required and decrease the number of available inputs per module by a factor of two. In the application shown in Figure 4-8, three different sources of 120 vac power are used to turn on three separate motor starters for Pumps 1, 2, and 3. This is a typical application for isolated ac output modules.

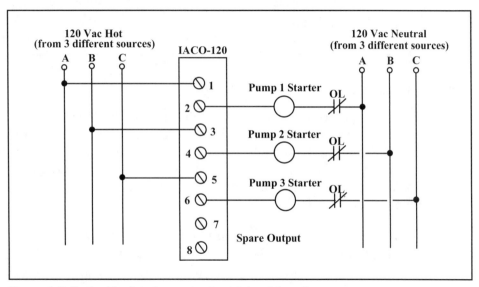

Figure 4-8. Typical isolated ac output module wiring diagram.

Analog I/O Modules

The availability of low-cost integrated circuits and industrial-grade electronic circuits greatly increased the capabilities of analog circuits in programmable logic controllers. This expanded capability led to the introduction of sophisticated analog input/output (I/O) modules.

Analog input modules allow measured quantities to be received from process instruments and other devices that provide analog data. Similarly, analog output modules make it possible to control devices that require a continuous analog signal. The analog I/O will allow monitoring and control of analog voltages and currents which are compatible with many sensors, motor drives, and process instruments. Using analog and special-purpose I/O makes it possible to measure or control most process variables as long as appropriate interfacing is used. Table 4-4 lists I/O devices that are typically interfaced to programmable logic controllers using analog modules.

Table 4-4. Typical Analog I/O Field Devices

Analog Input Devices	Analog Output Devices
Flow transmitters	Electric motor drives
Pressure transmitters	Analog meters
Temperature transmitters	Chart recorders
Analytical transmitters	Process controllers
Position transmitters	Current-to-pneumatic transducers
Potentiometers	Electrical-operated valve
Level transmitters	Variable-speed drives
Speed instruments	

Analog Input Modules

The analog input interface contains the circuitry needed to accept analog voltage or current signals from field devices. The voltage or current inputs are converted from an analog into a digital value by an analog-to-digital converter (ADC). The conversion value, which is proportional to the analog signal, is passed through to the controller's data bus and stored in a memory location for later use.

Typically, analog input interfaces have a very high input impedance, which allows them to interface field devices without signal loading. The input line from the analog device generally uses shielded, twisted-pair conductors. The shielded cable greatly reduces the electrical interferences from outside sources. The input stage of the interface provides filtering and isolation circuits to protect the module from additional field noise. A typical analog input connection is illustrated in Figure 4-9. In the example shown, the analog input module is providing the dc voltage required by the field current transmitters.

Most analog modules are designed to sense up to sixteen single-ended or eight differential analog input signals, representing flow, pressure, level, and the like. They then convert them into a proportional ten- to fifteen-bit binary word in memory. Inputs to a particular module must, in general, be all single-ended or all differential, and the type of signal can be selected via either hardware or software. The converted signals are stored in memory in the module and are sent to the processor memory in groups or blocks of data.

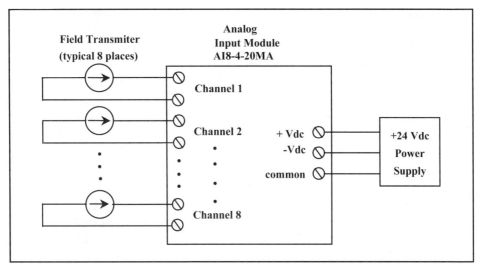

Figure 4-9. Typical eight-channel analog input module wiring diagram.

The control program uses configuration data to set up the analog module. Typical configuration information includes range selection (i.e., +1 to +5 vdc, 4 to 20 mA, etc.) and signal scaling.

Analog Output Modules

The analog output modules receive data from the PLC's central processing unit. The data is translated into a proportional voltage or current to control an analog field device. The digital data is passed through a digital-to-analog converter (DAC) and sent out in analog form. Isolation between the output circuit and the logic circuit is generally provided through optical couplers. These output modules normally require an external power supply with certain current and voltage requirements.

Special-Purpose Modules

A wide variety of special-purpose I/O modules are used in PLC systems. One PLC manufacturer has over 120 different types of I/O modules. We will discuss two of the more common special-purpose modules: the encoder/counter input module and the high-speed pulse input module.

Encoder/Counter Input Module

The encoder/counter input module provides a high-speed counter external to the processor that responds to input pulses sensed at the interface. This counter normally operates independently of either program scan or I/O scan. The reason for this is fairly simple: if the counter were dependent upon the PLC program, high-speed pulses would be missed during a program scan. Typical applications of the encoder/counter interface are operations that require direct encoder input to a counter that is capable of providing direct comparison outputs.

The encoder/counter input module accepts input pulses from an incremental encoder that provides pulses. These pulses signify position when the equipment rotates. The counter counts the pulses and sends them to the processor. Absolute encoders are generally used with interfaces that receive BCD or Gray code data, which represents the angular position of the mechanical shaft being measured.

During normal operations, the modules receive input pulses which are counted and then compared a preset value selected by the operator. The counter input module normally has an output signal that is energized when the input and preset counts are equal. However, this is not needed in most PLCs. Since the data is available in the CPU, the programmer can use a comparison function to drive an output in the control program.

The data communication between the encoder/counter interface and the CPU is bidirectional. The module accepts the preset count value and other control data from the CPU and transmits data and status to the PLC memory. The output controls are enabled from the control program that instructs the module to operate the outputs according to the count values received. The CPU, using the control program, enables and resets the counter operation.

Pulse Counter Input Modules

The pulse counter input modules are used to interface with field instruments that generate pulses, such as positive displacement (PD) flowmeters and turbine-type flowmeters. In a typical application, the flowmeter will generate a signal pulse with an amplitude of +5 volts depending on the volume of fluid passing through. Each pulse represents a fixed volume; for example, one pulse might equal one liter of fluid. In a typical application, the PLC counts the number of pulses received by the pulse input module and then calculates the total volume of fluid that passes for a fixed time period.

Intelligent I/O Modules

In the previous sections, we discussed discrete, analog, and special-purpose I/O modules. These will normally cover 90 percent of the I/O applications encountered in PLC systems. However, to process certain types of signals or data efficiently, microprocessor-based intelligent modules are required. These intelligent interfaces include those that condition input signals, such as thermocouple modules, or other signals that cannot be interfaced using standard I/O modules. These intelligent modules can perform complete processing functions, independent of the CPU and the control program scan. In this section, we will discuss two of the most commonly available intelligent modules: the thermocouple input and stepping motor output modules.

Thermocouple Input Modules

A thermocouple (T/C) input module is designed to accept inputs directly from a T/C, as shown in Figure 4-10. The T/C input module provides cold-junction temperature compensation that corrects for changes in ambient temperature around the T/C modules. This type of module operates much like the standard analog input, with the exception that low-level millivolt signals are accepted from the T/C (approximately 43 mV at maximum temperature for a J-type T/C). These signals are filtered, amplified, and digitized through an ADC. They are then sent to a built-in microprocessor, which linearizes the millivolt input signal so as to convert

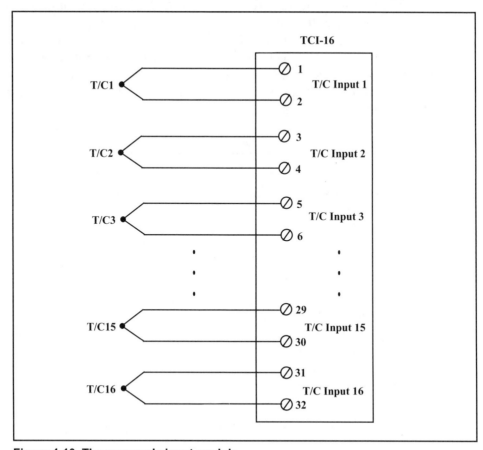

Figure 4-10. Thermocouple input module.

the signal into a temperature value. Finally, the temperature value is sent to the CPU on command from a program instruction. The temperature data is used by the PLC control program to perform temperature control and/or indication.

Stepping Motor Module

The stepping motor module generates a pulse train that is compatible with stepping motor translators (see Figure 4-11). The pulses sent to the translator normally represent distance, speed, and direction commands to the motor.

The stepping motor interface accepts position commands from the control program. Position is determined by the preset count of output pulses, by a forward- or reverse-direction command, and by the acceleration or deceleration command for ramping control, which in turn is determined by the rate of output pulses. These commands are generally specified during program control, and once the output interface is initialized by a start it will output the pulses according to the PLC program. Once the motion has started, the output module will generally not accept any commands from the CPU until the move is completed. Some modules may offer an override command that will reset the current position. It must be disabled to continue operation. The module also sends data regarding its status to the PLC processor.

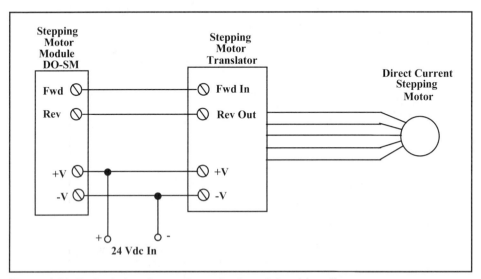

Figure 4-11. Typical stepping motor output module wiring diagram.

Communications Modules

There are six common types of communications modules used in PLC systems to communicate between system components. They are the ASCII module, the universal remote I/O link module, the serial communications module, the PCMCIA interface card, the Ethernet interface module, and the fiber-optic converter module. We will discuss each in turn.

ASCII Communications Module

The ASCII communications module is used to send and receive alphanumeric data between peripheral equipment and the controller. Typical peripheral devices with ASCII I/O include printers, digital display instruments, and so on. Depending on the manufacturer, this special I/O module is available with a communications circuitry interface that includes onboard memory and a dedicated microprocessor. The information exchange interface generally takes place via an RS-232C, RS-422, or RS-485 serial interface link or through a 20-mA dc current loop communications link.

The ASCII module will generally have its own RAM memory, which can store blocks of data that are to be transmitted. When the input data from the peripheral is received at the module, it is transferred to the PLC memory through a data transfer instruction. All the initial communication parameters, such as parity (even or odd) or nonparity, number of stop bits, and communications rate, can be selected via hardware or software.

Universal Remote I/O Link

The remote I/O link modules are used in larger programmable controller systems to allow I/O subsystems to be located some distance from the processor. The remote subsystem is used to interface with a unit process through a standard I/O rack and the required I/O modules. The rack will include a dc power supply to drive the internal circuitry of the I/O modules and a remote I/O adapter module to provide the communications with the processor unit. The I/O capacity of a single subsystem normally ranges from 32 to 256 points.

The subsystems are normally connected to the processor through a bus or star configuration. The distance from the processor to a given remote I/O rack normally ranges from one thousand feet to several miles, depending on the programmable controller type.

Remote I/O arrangements offer large cost savings on both wiring materials (wire and conduit) and labor for large control systems in which the field instrumentation is clustered at several remote process areas. If the processor is located in a main control room or some other central location, only the communication cable needs to be run between the processor and the field I/O racks instead of hundreds or thousands of field wires.

Remote I/O arrangements also have the advantage of allowing subsystems to be installed and tested independently, as well as permitting maintenance and troubleshooting on individual stations while other units continue to operate.

Serial Communications Module

The serial data communications module is normally used to communicate between the programmable controller and an intelligent instrument with a serial output, such as a weigh scale with a serial communications port. This serial communications module generally has two to four serial ports so it can be connected to RS-232, RS-422, and RS-485 standard communications interfaces.

PCMCIA Interface Card

In 1990, the Personal Computer Memory Card International Association (PCMCIA) developed a standard for a credit card-sized personal computer interface card. This standard defines an architecture and communications method for these PC interface cards. The interface cards developed under release 2.0 of the standard can be used for both data storage and I/O communications. PLC manufacturers developed PCMCIA cards so notebook personal computers could communicate with the PLC processor or data highway to perform PLC software and troubleshooting functions.

PCMCIA interface cards come with diagnostic software for verifying that the card is operating properly and for connecting it to the PLC communications network.

Ethernet Communications Modules

Ethernet interface modules are designed to allow a number of programmable controllers and other computer-based devices to communicate over a high-speed plant Ethernet communications network. This plant local area network (LAN) is able to transfer data and control information from one system to another at a high data transmission rate. Therefore, the control of an industrial facility can be distributed over a large number of programmable controllers, computers, and intelligent devices. In such a system, information is easily exchanged between control systems, but each system can independently control its part of the industrial plant. This greatly improves the reliability of the plant control system since sections of the plant can be down for modification or maintenance, while the remaining sections can continue to operate and produce product.

Fiber-Optic Converter Module

The fiber-optic converters transform electrical signals into light signals and transmit these signals through fiber-optic cables. At the other end of the cable, a second fiber-optic converter transforms light signals into electrical signals for use by the PLC system.

Designing I/O Systems

To correctly design I/O systems, the programmable controller manufacturer's specifications must be consulted and followed to prevent faulty operation or equipment damage. These specifications place limitations not only on the module but also on the field equipment it operates. The specifications fall into three categories: electrical, mechanical, and environmental.

Electrical Specifications

The typical electrical specifications for I/O modules include the following seven items: input voltage rating, input current rating, input threshold voltage, output voltage rating, output current rating, output power rating, and backplane current requirements.

The *input voltage (ac or dc) rating* specification lists the magnitude and type of signal the module will accept. In some cases, the specification states a range of input voltages instead of a fixed value. In this case, the maximum and minimum acceptable working voltages for continuous operation are listed. For example, the working voltage for a 120 vac input module might be listed as 95 to 135 vac.

The *input current rating* defines the minimum input current required at the module's rated voltage that the field device must be capable of supplying to operate the input module circuit.

The *input threshold voltage* is the voltage at which the input signal is recognized as being ON or TRUE. Some input modules also have an OFF voltage value at which the input is OFF or FALSE. For example, the ON voltage for TTL input module is defined as 2.8 vdc or greater, and the OFF voltage is defined as any voltage less than 0.8 vdc.

The *output voltage rating* specifies the magnitude and type of voltage source that can be controlled within a stated tolerance. An output module rated at + 24 vdc, for example, might have a working range of 20 to 28 vdc.

The *output current rating* defines the maximum current that a single output circuit in a module can safely carry under load. This current rating is normally specified as a function of the output circuit component's electrical and heat dissipation characteristics at an ambient temperature range (typically 0°C to 60°C). As the ambient temperature increases, the output current is normally derated. Exceeding the output current rating can result in a permanent short circuit or other damage to the output module.

The *output power rating* specifies the maximum total power that an output module can dissipate with all outputs energized. The output power rating for a single output is calculated by multiplying the output voltage rating by the output current rating. For example, if a 120-vac output module has a current rating of 2 amps, the power rating is $P = I \times V$ or $P = 2\,a \times 120\,v =$ 240 watts.

The *backplane current requirement* is the current demand that a particular I/O module internal circuitry places on the rack power supply. The system designer must add up the current requirements of all the installed modules in an I/O rack and compare the calculated value with the maximum current that can be supplied by the system power supply to determine if the power supplied is the correct size. If the rack power supply output is too low, the system will operate intermittently and deficiently. Some typical specifications for I/O modules are given in Table 4-5.

A sample problem will illustrate a typical I/O system design application.

EXAMPLE 4-1

Problem: Calculate the backplane current requirements for a programmable controller system with the following modules mounted in a PLC rack: five dc input modules, model number DCI-24; three dc output modules, model number DCO-24; two analog input modules, model number AI-4-20MA; and one analog output module mounted, model number AO-4-20MA. Assume that the rack power supply and the backplane power bus are both rated at 5 amps.

EXAMPLE 4-1 continued

Solution: Use the I/O specifications in Table 4-5 to find the total backplane current required, as follows:

       (a) DCI-24: 200 mA,

       (b) DCO-24: 200 mA,

       (c) AI-4-20MA: 400 mA, and

       (d) AO-4-20MA: 400 mA.

Thus, the total backplane current required is as follows:

     5 x 200 mA = 1000 mA

     3 x 200 mA = 600 mA

     2 x 400 mA = 800 mA

     1 x 400 mA = <u>400 mA</u>

               2800 mA

Therefore, the total backplane current demand is 2,800 mA or 2.8 amps. This is less than the rack power supply and backplane current rating of 5 amps, so I/O system current demand is acceptable. There is also extra current capacity for handling some spare modules in the future.

Table 4-5. Typical Specifications for I/O Modules

Model Number	Module Description	Backplane Current	I/O Power Rating
AI-4-20MA	Analog input (4-20 mA)	400 mA	100 mW
AO-4-20MA	Analog output (4-20 mA)	400 mA	100 mW
ACI-120	120 vac input	200 mA	240 W
ACI-120	220 vac input	200 mA	440 W
DCI-12	12 vdc input	200 mA	24 W
DCI-24	24 vdc input	200 mA	48 W
DCI-48	48 vdc input	200 mA	96 W
IACI-120	Isolated ac input	200 mA	240 W
ACO-120	120 vac output	250 mA	240 W
ACO-220	220 vac input	250 mA	440 W
DCO-12	12 vdc output	200 mA	24 W
DCO-24	24 vdc output	200 mA	48 W
DCO-48	48 vdc output	200 mA	96 W
CC-4NO	4 NO contact output	500 mA	240 W
TTLI	TTL input	150 mA	100 mW
TTLO	TTL output	150 mA	100 mW
IACO-120	Isolated ac output	250 mA	240 mW

Mechanical and Environmental Specifications

The typical mechanical specifications are I/O-points-per-module and wire size. The I/O-points-per-module specification simply defines the number of field points that are controlled or sensed by a module. Typically, modules will have two, four, eight, sixteen, or thirty-two I/O points per module. The higher-density modules require higher operating current, so the backplane current must be carefully checked. The number of wires is also increased and it might be a problem for larger gage wires. The wire-size specification defines the number of conductors and the largest wire gauge that the I/O terminal points will accept. For example, a typical wire specification for an eight-point input or output ac module is 2-14 AWG wires per terminal. However, higher-density modules (i.e., sixteen- and thirty-two-point modules) might only allow a single 14 AWG wire per terminal.

Environmental Specifications

The important environmental specifications are the ambient temperature rating and the humidity rating. The ambient temperature specification defines the maximum temperature of the surrounding air in which the input/output system will operate properly. This rating is based on the heat-dissipation characteristics of the circuit components inside the I/O modules, which are much higher than the module ambient temperature rating. A typical value of ambient temperature rating is 0 °C to 40 °C. Exceeding the ambient temperature rating can be dangerous because the internal circuits of the modules will act erratically. This will result in undesirable outputs to the process being controlled. Exceeding the ambient temperature rating will also decrease the life of the module.

The humidity specification is typically 5 percent to 95 percent without condensation. The system designer must ensure that the humidity is properly controlled in the control panel where the I/O system is installed.

EXERCISES

4.1   Describe the main purpose of programmable controller I/O systems.

4.2   What is the backplane on a typical PLC I/O equipment rack?

4.3   What is a discrete input signal? Give five typical examples of discrete-input field devices.

4.4   Give five examples of typical discrete-output field devices.

4.5   What is the purpose of the bridge rectifier in the ac input circuit shown in Figure 4-2?

4.6   What is the purpose of the optical coupler in the ac input circuit shown in Figure 4-2?

4.7   Describe how the internal circuits in the typical ac output module shown in Figure 4-5 operate.

4.8   Draw a wiring diagram for a typical +24 vdc input module to which are connected four temperature-high switches (TSH-100, 101, 102, and 103) and three pressure-low switches (PSL-210, 211 and 212).

4.9   Draw a wiring diagram for a typical +120 vac output module connected to four solenoid valves (TV-100, 101, 102, and 103) and three pump starters (P-100, 200 and 300) with overload switches.

4.10   Using the backplane current values listed in Table 4-5, calculate the backplane current requirements for a sixteen-slot I/ O rack in which the following modules are installed: (a) four 12 vdc input modules (model number DCI-12), (b) six 12-vdc output modules (model number DCO-12), (c) four analog input modules (model number AI-4-20MA), and (d) two analog output modules (model number AO-4-20MA).

BIBLIOGRAPHY

1.   Allen-Bradley Co., Inc. *Allen-Bradley Automation Systems* (Allen-Bradley, December 1994).

2.   Allen-Bradley Co., Inc. *Allen-Bradley Industrial Computer and Communications Group Product Guide* (Allen-Bradley, 1987).

3.   Allen-Bradley Co., Inc. *Processor Manual PLC-5 Family Programmable Controllers* (Allen-Bradley, 1987).

4.   Bryan, L. A., and E. A Bryan. *Programmable Controllers: Theory and Implementation* (Industrial Text Co., 1988).

5.   Hughes, T. A. *Measurement and Control Basics, 2nd ed.* (ISA, 1995).

6.   Jones, C. T., and L. A. Bryan. *Programmable Controllers: Concepts and Applications* (International Programmable Controls, 1983).

5

Memory and Addressing

Introduction

Programmable controller systems store information and programs in memory. The information stored there determines how the programmable controller will process input and output data. In this chapter, we will discuss the components and structure of memory, types of memory, memory organization, addressing memory, and I/O addressing. Finally, we will focus on the PLC hardware-to-software interface.

Memory Components and Structure

Programmable controller memories can be visualized as a two-dimensional array of storage cells, each of which can store a single bit of information in the form of a one or a zero. This single binary digit or "bit" gets its name from the first two letters of the word *binary* and the last letter of *digit*. A bit is the smallest structural unit of memory and stores information in the form of ones and zeros, though they are not actually in each cell. Each cell has a voltage that is present at the output of an electronic circuit, which is indicated by a one, or a voltage close to zero, which is indicated by a zero.

The bit is set or ON if the stored bit is one, and OFF if the stored bit is zero. In most cases, it is necessary for the processor to handle more than a single bit. For example, when the processor is transferring data to and from memory, storing numbers, and programming codes, it needs a group of bits called a *byte* or *word*. A byte is defined as the smallest group of bits that can be handled by the CPU at one time. In programmable controllers, byte size is normally eight bits and word size is normally two bytes or sixteen bits. However, the word size can be smaller or larger depending on the specific microprocessor being used.

Memory capacity is specified in thousands or "K" increments, where 1K is 1,024 words (i.e., 2^{10} = 1,024) of storage space in most cases. Programmable controller memory capacity may vary from less than 1,000 words to over 64,000 words (64K words), depending on the programmable controller manufacturer. The complexity of the control plan, the number of I/O points, and type of I/O will determine the amount of memory required.

Word length is usually two bytes (sixteen bits) or more in length. Typical word lengths in programmable controllers are eight, sixteen, or thirty-two bits. A sixteen-bit word is shown in Figure 5-1.

Some programmable controllers use the octal numbering system to identify each bit in a given word, as shown in Figure 5-1. The most significant bit (MSB) is bit 17, and the least significant bit (LSB) is bit 00.

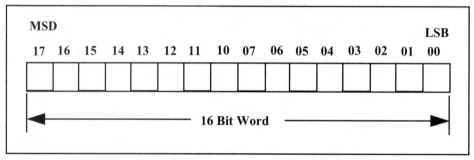

Figure 5-1. Basic memory word.

A simple sixty-four-cell memory array is shown in Figure 5-2. This array consists of eight rows and eight columns. This sixty-four-bit array requires only six bits to address a given cell. A cell is normally an electronic circuit, called a *flip-flop*, that can have a value of +5 volts (logic 1) or 0 volts (logic 0). To retrieve data from the memory array, row and column address decoders select the appropriate cell.

These memory arrays are normally provided by integrated circuits (IC). A typical IC unit contains many thousands of memory cells arranged in various ways. An 8K-bit (8,096 cells) memory integrated circuit, for example, could be arranged as 8K memory cells of one bit each, or 1K bytes of eight bits each. The number of groups (bits, bytes, or words) addressed is a function of 2^n, for example 1K = 2^{10}, 4K = 2^{12}, 8K = 2^{13}, and so on. The value n is the number of address bits needed to select each

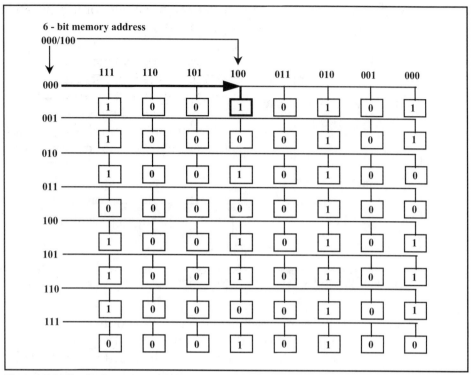

Figure 5-2. Typical memory array.

separate group. For one thousand words, it is necessary to use ten bits to address each word of storage, with the word size being eight bits, sixteen bits, and thirty-two bits. For a 1K × 8 memory, the IC would require ten address bits to select 1K words of memory. A typical 1K × 8-bit word is shown in Figure 5-3. The IC has eight pins for input and output data bits, ten connection pins for selecting addresses, two pins for control signals (chip-enable and read/write), and two pins for dc power. The two power supply pins are used for +5 vdc and ground. The read/write (R/W) control signal is used to determine whether the data bits are read into memory (R/W signal low) or the data is sent out from memory (R/W is high). The chip-enable control signal is used to select the operation of each separate chip when a group of integrated circuits is used to provide a larger memory than is provided by only one chip.

Memory Types

This section will discuss the types of memory that are generally used in programmable controllers as well as their application to the type of data or information stored. In selecting which memory to use, a system designer is concerned with volatility and ease of programming. He or she is concerned with volatility because memory holds the process control

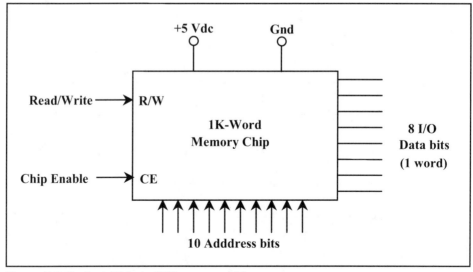

Figure 5-3. Typical 1K-byte R/W memory chip.

program, and if this program is lost a plant's production will be down. Since the memory is involved in any interaction that takes place between the user and the PLC it's important that it be easy to alter. This interaction begins with the initial system programming and debugging and continues with on-line changes, such as changing timer and counter preset values.

Read/Write (R/W) Memory

Read/write (R/W) memory is designed so that data or information can be written into or read from any unique location. Figure 5-4 shows that data can be placed into R/W memory using the write mode and it can be retrieved from R/W memory using the read mode. The address input to the R/W memory specifies the location or the address of the data to be read or the location to be written into.

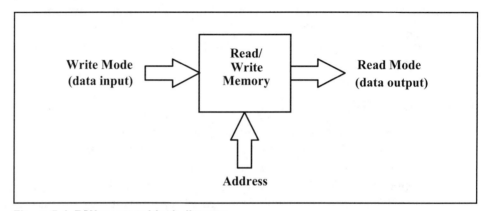

Figure 5-4. R/W memory block diagram.

Programmable controllers, for the most part, use R/W memory with battery backup for application memory. R/W memory provides an excellent way to easily create and alter a control program as well as allowing for data entry. In comparison to some other memory types, R/W memory is relatively fast. The only important disadvantage of battery-supported R/W memory is that it requires a battery that might fail at a critical time. However, most programmable controllers have battery-low lights to alert operations personnel to replace the memory power backup battery.

R/W memory is normally an integrated circuit chip (see Figure 5-3) that stores individual bits of data in multiple rows and columns of cells. This row-and-column arrangement allows each cell to have a unique designation, called an *address*. This address consists of a row identifier and a column identifier, both of which are expressed as binary numbers.

Read-Only Memory (ROM)

Read-only memory is designed to permanently store a fixed program that normally cannot or will not be changed. It gets its name from the fact that its contents can be read but not written into or altered once the data or program has been stored. Figure 5-5 shows that data can be used only in the read mode. As with R/W memory, ROM has an address input where the location of the data to be read can be specified. Because of their design, ROMs are generally immune to changes caused by electrical noise or loss of power. A PLC's executive or operating system program is normally stored in ROM.

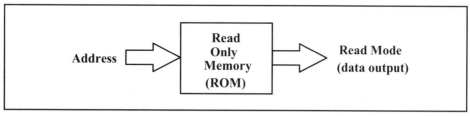

Figure 5-5. ROM memory block diagram.

Programmable controllers rarely use ROM for the control applications' program memory. However, in applications that require fixed data, ROM offers advantages where speed, cost, and reliability are factors. Generally, ROM-based PLC programs are produced at the factory by the equipment manufacturer. Once the original set of instructions is programmed, the user can never alter it. The manufacturer will write and debug the program using a read/write-based controller or computer, and then the final program is entered into R/W or read-only memory. Read-only memory is also found in the application memory of dedicated programmable controllers systems such as microwave ovens, vending machines, clothes washers, and the like.

Programmable Read-only Memory (PROM)

The programmable read-only memory (PROM) is a special type of ROM that is rarely used in most programmable controller applications. However, when it is used, it will most likely be a permanent storage to some type of random access memory (RAM). PROM is programmable and, like other ROM, has the advantage of nonvolatility. However, its disadvantages include the special programming equipment it requires and the fact that, once programmed, it cannot be erased or altered. Any program change would require a new set of PROM chips. PROM memory might be suitable for storing a program that has been thoroughly checked while stored in RAM and will not require further changes or on-line data entry.

Electrically Erasable Programmable ROM (EEPROM)

Electrically erasable programmable ROM (EEPROM) is a special type of PROM that can be reprogrammed after it is completely erased using an electrical method. The EEPROM can be considered a temporary storage device in that it stores a program until it is ready to be changed. EEPROM provides an excellent storage medium for a control program that requires nonvolatility but not program changes. Many manufacturers of equipment with built-in PLCs use EEPROM-type memories to provide permanent storage of the machine program after it has been developed and debugged and is fully operational.

A control program composed of EEPROM alone would be unsuitable if on-line changes or data entries are a requirement. However, many PLCs offer EEPROM control program memory as an optional backup to battery-supported RAM. EEPROM makes a suitable memory system because it combines permanent storage capability and easily altered read/write memory.

Memory Organization

A typical programmable logic controller memory has two main parts: the *system memory* and the *application memory*. As shown in Figure 5-6, the system memory is a permanently stored collection of programs and registers that consists of the operating system program, diagnostics software, and system status registers. The operating system directs activities such as executing the control program, communicating with the peripheral devices, and other system housekeeping functions. The

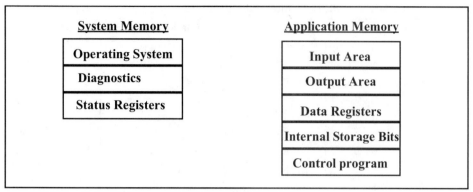

Figure 5-6. Typical PLC memory organization.

application memory consists of the input area, output area, data or information storage registers, internal storage bit area, and the control program.

The system memory and the application memory have different storage and retrieval requirements. The system memory contains the instructions to operate the CPU, a set of diagnostics programs, and status registers. The application memory contains the input image area (designated by an *I* in most PLCs), the output image area (designated by an *O* or *Q* in most PLCs), the control program, and the data registers. They also use different types of memory. The operating system portion of the system memory requires a memory that permanently stores its contents and cannot be deliberately or accidentally altered by the loss of electrical power or by the user. Therefore, some type of ROM would be used. On the other hand, the user would need to alter the control program or the input/output data for any given application, so R/W memory is used in the input/output data and the control program sections of memory.

File Structure

A *memory file* is defined as a group of words in memory that has a dedicated function. The input and output files are most common type of files in a PLC. They are the memory words set aside for the external input and output bits assigned to the physical I/O points in a PLC system. The file structure varies with the PLC manufacturer. The file structure for Allen-Bradley PLC5s is listed in Table 5-1. There are nine file types listed, but additional files can be assigned as needed.

File number 0 is for the output image table, and it has the file letter *O*. File 1 is the input image table area of memory, and it has the file letter *I*. File 2 is reserved for the status bits and words in the PLC. Input and output image tables and the status file areas can have a maximum of thirty-two elements or words.

Table 5-1. A-B PLC5 Memory File Structure

File Number	File Letter	File Name	Maximum Words
0	O	Output image	32
1	I	Input image	32
2	S	Status	32
3	B	Bits for program use	1000
4	T	Timers	1000
5	C	Counters	1000
6	R	Control registers	1000
7	N	Integer	1000
8	F	Floating point	1000

For comparison, Table 5-2 gives a partial listing of memory file structure for Siemens Simatic S7 PLCs. In both Allen-Bradley and Siemens PLCs the input image bits are implemented with an *I* in programming. The output image bits are designated by an *O* in Allen-Bradley PLC5s and a *Q* in Siemens S7 PLCs.

Table 5-2. File Structure for Siemens Simatic S7 PLCs

File Letter	File Type	Description	Max. Address Range
I	Input image	Input bit	0.0 to 65535.7
IB		Input byte	0 to 65535
IW		Input word	0 to 65534
ID		Input double word	0 to 65532
Q	Output image	Output bit	0.0 to 65535.7
QB		Output byte	0 to 65535
QW		Output word	0 to 65534
QD		Output double word	0 to 65532
M	Bit memory	Memory bit	0.0 to 255.7
MB		Memory byte	0 to 255
MW		Memory word	0 to 254
MD		Memory double word	0 to 252
PIB	External inputs	Periph. input byte	0 to 65535
PIW		Periph. input word	0 to 65534
PID		Periph. In double word	0 to 65532
PQB	External outputs	Periph. input byte	0 to 65535
PQW		Periph. input word	0 to 65534
PQD		Periph. Out double word	0 to 65532
T	Timers	Timer	0 to 255
C	Counters	Counter	0 to 255
DBX	Data block	Data bit	0.0 to 65535.7
DBB		Data byte	0 to 65535
DBW		Data word	0 to 65534
DBD		Data double word	0 to 65532

Internal Storage Bits

Most programmable controllers assign an area for internal storage bits. These storage bits are also called *internal outputs*, *internal coils*, or *internal control bits*. The internal output operates just like any output that is controlled by programmed logic. However, the output is used strictly for internal logic programming and does not directly control an output to the process. Internal outputs are used to interlock logic in the control program.

Internal outputs include the "done" bits on counters and timers as well as internal logic bits of various types. Each internal output bit is referenced by an address in the control program and has a storage bit of the same address. When the control logic is TRUE, the internal (output) storage bit turns ON.

User Program Memory Area

The user program memory area of the application memory is used to store the process control logic program. All of the controller instructions that control the machine or process are stored here. The addresses of the real and internal I/O bits are specified in this section of memory as well. When the PLC is in the run mode and the control program is executed, the CPU interprets these memory locations and controls the bits in the data table, which correspond to a real or internal I/O bit. The control program is interpreted when the processor executes the control program.

The maximum amount of user program memory that is available is normally a function of the controller memory size. In medium and large programmable controllers, the user program size is normally flexible. It can be altered by changing the data table size so it meets the minimum data storage requirements. In small PLCs, however, the user program size is normally fixed.

Application Memory Size

The size of the application memory is an important factor in the design of programmable controller-based control systems. Specifying the correct memory size can reduce hardware costs and avoid lost time later. If you calculate memory size properly you won't purchase a PLC that lacks adequate capacity or expandability.

Application memory size may be expandable to some maximum point in some controllers, but it is not expandable in some of the smaller PLCs. (Smaller PLCs are defined in this context as units that control ten to sixty-four input/output devices.) Programmable controllers that handle sixty-four or more I/O devices are usually expandable in increments of 1K, 2K, 4K, and so on. (Recall that each *K* represents 1,024 or 2^{10} word locations in memory.) The memory in the larger controllers is generally 64K or higher.

The stated memory size of a PLC gives only a rough indication of the memory space available to the user since some of the memory is used by the controllers for internal functions.

The main obstacle in determining memory size for an application is that the complexity of the control program is not determined until after the equipment is purchased. However, we generally know the number of I/O points in the system before the hardware is procured. The system designer can therefore estimate application memory size by multiplying the number of I/O points by twenty words of memory. For example, if the system has one hundred I/O points, the program will generally be equal to or less than two thousand words. Keep in mind that program size is affected by the sophistication of the control program. If the application requires data handling or complex control algorithms, such as PID control, then additional memory will be required.

After the system designer determines the minimum memory required for an application, he or she will normally add an additional 25 percent to 50 percent for program changes, modification, or future expansion.

EXAMPLE 5-1

Problem: Determine the memory size needed for a programmable controller system that has 175 input points and 125 output points. Assume that 25 percent spare memory capacity is required for the system.

Solution: Total I/O points = 175 + 125 = 300 points

Memory required = 300 points x 20 words/point + 25%

= (6000 + 6000 x 0.25) words

= (6000 + 1500) words

= 7.5 K words

I/O Addressing

Since one of the main purposes of a PLC is to control the inputs and outputs (I/O) of field devices, these I/O must occupy a location in the processor memory where they can be addressed in the PLC control program. Each terminal on an input or output module that can be wired to a field device occupies a bit within PLC memory. The part of memory that houses I/O addresses is called the *input image table* and *output image table*.

Input Image Table

The input image table is an array of bits that stores the status of discrete inputs from the process, which are connected to inputs. The number of bits in the table is equal to the maximum number of inputs. A controller with a maximum capacity of sixty-four inputs would require an input table of sixty-four bits. Each connected input has a bit in the input table that corresponds exactly to the terminal to which the input is connected. If the input is ON, its corresponding bit in memory is ON or logic 1. In most programmable controllers, the input is ON if voltage is present on the input terminal. If no voltage is present, the corresponding bit is cleared or turned OFF or set to logic 0. The input table is continuously being changed to reflect the current status of the connected input devices. The control program uses this status information to determine the TRUE or FALSE state of the program instructions.

Figure 5-7 shows a typical example of a single input bit in an input image table. In this example, input point I:007/12 is identified on the memory map.

Output Image Table

The output image table is an array of bits that controls the status of discrete output devices, which are connected to output interface circuits.

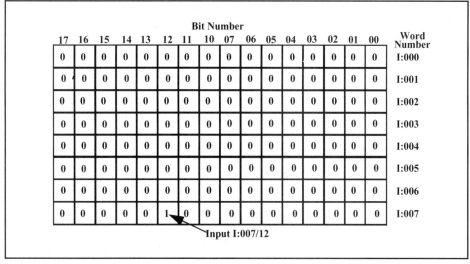

Figure 5-7. Input bit example in an Allen-Bradley PLC5 input image table.

Each connected output has a bit in the output image table that corresponds exactly to the terminal to which the output is connected. The bits in the output table are controlled by the PLC processor as it interprets the control program and are updated accordingly during the I/O scan. If a bit is turned ON or logic 1, then the connected output electrical circuit is activated, and there is a voltage present on the output terminal. If a bit is cleared or turned OFF or logic 0, the output is switched off.

Figure 5-8 shows a typical example of a single output bit in an output image table. In this example, output bit location O:017/16 is shown on the output memory map. The letter O indicates an output, the word address is 017, and the bit is 16 or the last bit in the memory word.

	Bit Number															Word Number
17	16	15	14	13	12	11	10	07	06	05	04	03	02	01	00	
0	1	1	0	0	0	1	0	0	1	0	0	0	0	1	0	O:000
0	1	0	0	0	1	1	0	0	1	1	0	0	0	1	0	O:001
0	1	1	0	0	0	1	0	0	0	1	0	0	0	0	0	O:002
0	0	1	0	0	0	1	1	0	0	1	0	1	0	1	0	O:003
0	1	1	0	1	0	1	0	0	1	1	0	0	0	1	1	O:004
0	1	0	0	0	1	1	0	0	0	1	0	0	0	0	0	O:005
1	1	0	0	0	1	0	1	0	1	1	0	0	0	1	0	O:006
0	1	1	0	0	0	1	0	0	0	1	0	0	1	1	0	O:007

Output O:007/02

Figure 5-8. Output bit example in an Allen-Bradley PLC5 output image table.

Hardware-to-Software Interface

Probably the most important thing to understand about programmable controllers is how the processor uses process inputs, which are sensed by the input circuits, to activate output devices in order to control the process. This hardware-to-software interface, which we briefly discussed earlier, occurs in the input/output image tables. The instruction address is what connects the software control program to the hardware input and output terminations.

The addressing of the PLC inputs and outputs connects the physical location of an I/O module terminal to a memory bit location. The structure or density of an I/O module, that is, eight-point, sixteen-point, or thirty-two-point, is directly related to the bits the module occupies in

PLC memory. For example, in the eight-point input module shown in Figure 5-9, the eight bits 00 through 07 occupy eight positions in an input memory table in the processor memory.

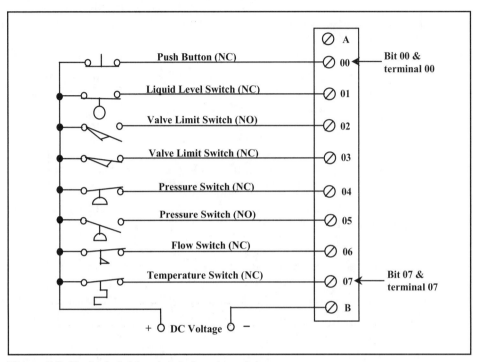

Figure 5-9. Typical eight-point dc input module.

We will next present several typical I/O addressing methods used in programmable controller systems. The first addressing scheme is used in the Allen-Bradley (A-B) PLC5 family of programmable logic controllers.

Allen Bradley PLC5 Discrete I/O Addressing

The A-B PLC5 discrete I/O addressing method uses a six-position code (a:bbc/dd) to reference both an I/O memory address and a hardware physical location. In this system, the left-most position (*a*) is the letter *I* for a discrete input and the letter *O* for a discrete output. The next two digits (*bb*) are the rack number. The next digit (*c*) is the I/O group number (0-7). The remaining two digits represent the input or output bit or terminal numbers, 00-07 or 10-17.

For example, the address I:001/07 indicates an input device connected to rack 00 and I/O group number 1 at terminal 07 and memory bit 07. The address O:074/10 indicates an output device connected to rack 07 at terminal 10 in I/O group 4. Example 5-2 will illustrate the Allen-Bradley PLC5 addressing scheme.

EXAMPLE 5-2

Problem: List the memory bit address for a discrete signal input connected to an A-B PLC5/ 15 rack 03, I/O group number 4, terminal 10.

Solution: I:034/10

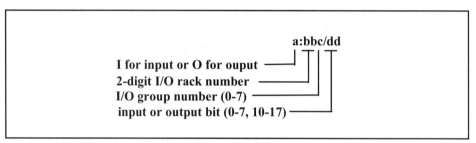

Figure 5-10. A-B PLC5 discrete I/O addressing scheme.

A hardware-to-software interface for an Allen-Bradley PLC5 application is illustrated in Figure 5-11. This application shows the operational relationship between the field devices, the discrete input/output memory bits, and the user ladder logic program.

In the example shown in Figure 5-11, if the high-level switch connected to an input module in rack 0, I/O group 0, and terminal 7 is closed, the internal software bit I:000/07 will be set to 1. The dotted line from terminal 7 to memory bit location I:000/07 indicates a connection within the PLC5 system. Imagine that at the same time the valve-open position switch that is connected to terminal 13 of the same input module is closed. In that case, input bit I:000/13 will be set to 1, and the ladder logic rung will have logic continuity, and output bit O:001/03 will be set to 1. This will energize the solenoid valve connected to terminal 3 of rack 0 and I/O group 1.

The ladder logic rung at the bottom of Figure 5-11 shows an example of two external input bits being used to set an external output bit. We can also have ladder logic rungs that use internal bits to set both external and internal bits. In some cases, the logic in the rung will use a combination of both internal and external bits to control an output bit.

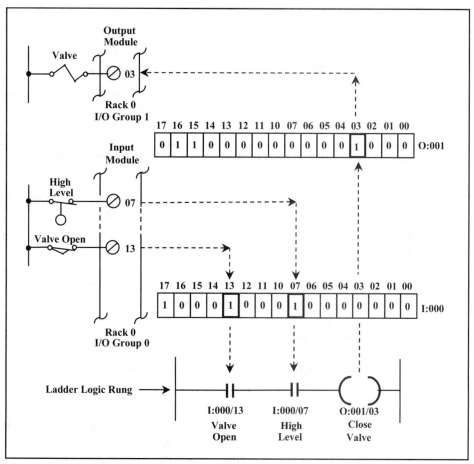

Figure 5-11. Hardware-to-software interface diagram.

Siemens Simatic S7-300 Discrete I/O Addressing

In this section, we will discuss the discrete I/O addressing for the Siemens Simatic S7 family of programmable controllers. This will enable us to compare and contrast them to the Allen-Bradley PLC5 discrete I/O addressing.

Figure 5-12 shows an eleven-slot equipment rack for a Siemens Simatic S7-300 PLC. The slot numbers in the rack influence the addressing scheme for the S7-300 family of PLCs. The first I/O address in the module is determined by its location in the rack. The first slot is reserved for the rack power supply, and of course no I/O addresses are needed for the power supply. The CPU must be located next to the power supply in slot number 2, and no I/O addresses are assigned to the CPU. Slot number 3 in Figure 5-12 contains a communications interface module (CIM). This module is used to connect the CPU in the main I/O rack with the I/O in the expansion racks. No I/O addresses are assigned to the interface module in

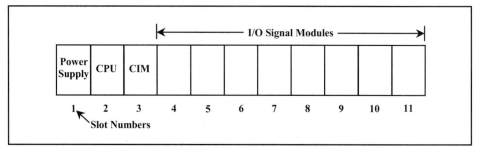

Figure 5-12. Siemens S7-300 rack arrangement and slot assignments.

slot 3. Even if no interface module is installed in slot 3, it cannot be used for other I/O modules. In other words, slot 3 in the Simatic S7 is logically reserved in the CPU for communications interface modules. Slot number 4 is the first slot used for I/O signal modules.

The relationship between physical rack and slot position and the I/O module position is shown in Figure 5-13. The discrete addresses for both inputs and outputs begin with word address 0 and bit address 0 (0.0) in slot 3 of rack 0. They continue to word 95 and bit 7 in slot 11 of rack 2. Each discrete I/O module is allocated four bytes (thirty-two bits) of a memory word addresses, regardless of the actual I/O point count of the module.

The Siemens discrete I/O addressing method uses a three-position code (abb.cc) to reference an I/O memory address. In this system, the left-most position (a) is the letter I for a discrete input and the letter Q for a discrete output. The next two digits (bb) are the memory byte number assigned to the I/O slot, and the digit (c) to the right of the dot is the I/O image table bit number (0 through 7).

For example, if the I/O module in slot 5 in rack 0 is a sixteen discrete input module, the first eight inputs would be assigned the addresses of 4.0 to 4.7. The second eight inputs would be assigned the bit addresses of 5.0 to 5.7, as shown in Figure 5-13.

Example 5-3 is a typical example of the addressing method used for Siemens S7-300 programmable controllers.

EXAMPLE 5-3

Problem: List the input bit addresses available for use on a thirty-two-bit discrete input module installed into slot 10 of rack 2 on the Siemens S7-300 system shown in Figure 5-13.

Solution: Discrete input can be connected to I/O points: 88.0 to 88.7, 89.0 to 89.7, 90.0 to 90.7, and 91.0 to 91.7.

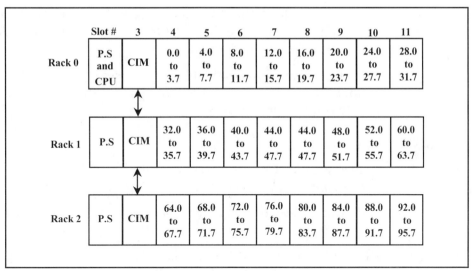

Figure 5-13. Typical Siemens S7-300 discrete I/O address configuration.

EXERCISES

5.1 List the most common types of memory used in programmable controller applications.

5.2 Describe what read/write memory is and give a typical application for it.

5.3 What is the main purpose of read-only memory in PLC applications?

5.4 Describe the function and purpose of the two main parts of programmable controller memory.

5.5 What are the input and output image tables used for in programmable controllers?

5.6 What is the memory size needed for a programmable controller system with 235 input points and 195 outputs? Assume that 25 percent spare memory capacity is required in the application.

5.7 What is the memory bit address for a discrete field device connected in rack 1 at I/O group 3 and terminal 1 of an Allen-Bradley PLC5 input module?

5.8 What is the memory bit address for a discrete field device connected to rack 2 at module group 1 and terminal 7 of an Allen-Bradley PLC5 output module?

5.9 What is the memory bit address for a discrete field device connected to rack 2 at I/O group 2 and terminal 13 of an Allen-Bradley PLC5 output module?

5.10   List the bit addresses available for use by a sixteen-bit discrete output module installed in slot 4 of rack 1 of the Siemens S7-300 system shown in Figure 5-13.

5.11   List the bit addresses available for use by a sixteen-bit discrete input module installed in slot 5 of rack 2 of the Siemens S7-300 system shown in Figure 5-13.

BIBLIOGRAPHY

1.   Allen-Bradley Co., Inc. *Processor Manual PLC-5 Family Programmable Controllers* (Allen-Bradley, 1987).

2.   Allen-Bradley Co., Inc. *Programming and Operations Manual, PLC-2/30 Programmable Controllers* (Allen-Bradley, 1988).

3.   Boylestad, R. L., and L. Nashelsky. *Electronic Devices and Circuit Theory*, 3d ed. (Prentice-Hall, 1982).

4.   Gilbert, R. A., and J. A. Llewellyn. *Programmable Controllers: Practices and Concepts* (Industrial Training Corp., 1985).

5.   Hughes, T. A. *Measurement and Control Basics*, 2nd ed. (ISA, 1995).

6.   Jones, C. T., and L. A. Bryan. *Programmable Controllers: Concepts and Applications* (International Programmable Controls, 1983).

7.   Time-Life Books, Inc. *Understanding Computers: Memory and Storage* (Time-Life Books, 1987).

6

Ladder Diagram Programming

Introduction

Programming languages allow the user to communicate with the programmable logic controller (PLC) via a programming device. PLC manufacturers use several different programming languages, but they all use instructions to convey a basic control plan to the system.

A program is written by combining instructions in a certain order. Rules govern the way in which the instructions are combined and the actual form the instructions take. These rules and instructions constitute a programming language. The three most common types of languages encountered in programmable controller systems are as follows:

1. Ladder diagram (LAD)

2. Statement list (STL)

3. Function block diagram (FBD)

The ladder diagram (LAD) is the most common programmable controller language, so we will address it first in this chapter. It consists of a set of instructions that will perform the most basic type of control functions: relay-type logic, timing and counting, and basic math operations. However, depending on the programmable controller model, the instruction set may be extended or enhanced to perform other operations. These additional functions are used for analog control, data manipulation, reporting, complex control logic, and other functions.

Basic LAD Instruction Set

LAD language is a symbolic instruction set that is used to create a programmable controller program. It is composed of six categories of instructions: relay type, timer/counter, data manipulation, arithmetic, data transfer, and program control. The ladder instruction symbols used to obtain the control logic to be entered into memory.

The main function of the LAD program is to control outputs based on input conditions. This control is accomplished through the use of what is referred to as a *ladder rung*. A ladder logic rung consists of a set of input conditions that are represented by relay contact-type instructions and, at the end of the rung, an output instruction, which is represented by the relay coil symbol.

Coils and contacts are the basic symbols of the ladder diagram instruction set. The contact symbols programmed in a given rung represent conditions that need to be evaluated to determine how the output should be controlled. All discrete outputs are represented by coil symbols.

When programmed, each contact and coil is referenced with an address number that identifies what is being evaluated and what is being controlled. Recall that these address numbers reference the data table location of either an internal bit or a connected input or output. Regardless of whether a contact represents an input/output connection or an internal bit, it can be used throughout the program whenever that condition needs to be evaluated.

The format of the rung contacts is dependent on the desired control logic. Contacts may be placed in configurations: such as series, parallel, or series and parallel, that are required to control a given output. For an output to be activated or energized at least one left-to-right path of contacts must be closed. A complete closed path is referred to as having *logic continuity*. When logic continuity exists in at least one path, it is said that the rung condition is TRUE or ON. The rung condition is FALSE or OFF if no path has continuity.

In the early years, the standard ladder instruction set was limited to performing only relay-equivalent logic functions using the basic relay-type contact and coil symbols, like those illustrated in Figure 6-1. A need for greater flexibility, coupled with developments in technology, led to extended ladder diagram instructions that perform data manipulation, arithmetic, and program flow control. We will discuss relay-type instructions first and the more advanced instructions later in this chapter.

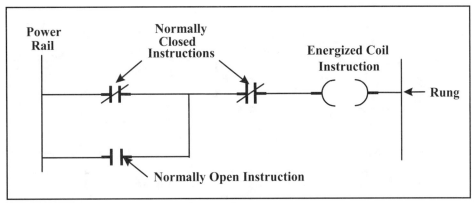

Figure 6-1. Basic ladder diagram instructions.

Relay-Type Instructions

The relay-type instructions are the most basic of programmable controller instructions. They provide the same capabilities as hardwired relay logic, discussed in chapter 1, but with greater flexibility. These instructions primarily provide the ability to examine the ON or OFF status of a specific bit addressed in memory and to control the state of an internal or external output bit. The following sections discuss the relay-type instructions that are most commonly available in any controller that has a ladder diagram instruction set.

Normally Open Instruction

The normally open (NO) instruction is programmed when the presence of the input signal is needed to turn an output ON. When evaluated, the referenced address is examined for an ON (logic 1) or an OFF (logic 0) condition. The referenced address may represent the status of an external input, an external output, or an internal output. If, when examined, the referenced address bit is ON or logic 1, then the normally open instruction will allow logic flow, as shown in Figure 6-2a. If the normally open instruction is OFF or logic 0, then the normally open instruction will assume its normal state of OPEN, thus breaking logic continuity and preventing logic flow, as shown in Figure 6-2b. As an aid in troubleshooting control programs, almost all PLC software programming packages will highlight logic bits to indicate the ON state of the input and output bits. Some programming packages will highlight the entire rung, as shown in Figure 6-2a, if the logic continuity or flow of the entire rung is activated or ON.

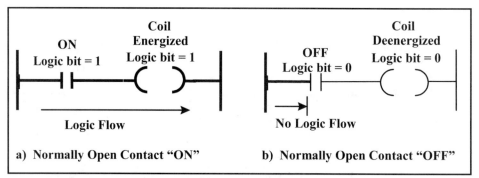

Figure 6-2. Logic flow for normally open contact instruction.

Normally Closed Instruction

The normally closed (NC) instruction is used when the absence of the referenced signal is needed to turn an output ON. When evaluated, the bit address of the NC instruction is examined for an ON (1) or an OFF (0) logic condition. The address referenced by the NC instruction may represent the status of an external input, an external output, or an internal output. If, when examined, the referenced bit address is OFF or logic bit 0, then the normally closed instruction will remain closed, allowing logic continuity, as shown in Figure 6-3a. If the referenced address is ON (logic bit 1), then the normally closed contact will open and break logic continuity or flow, as shown in Figure 6-3b.

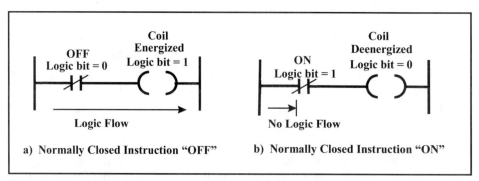

Figure 6-3. Logic flow for normally closed instruction.

Output Coil Instruction

The output coil or energize coil instruction is programmed to control either an output connected to the controller or an internal output bit. The output coil instruction bit is designated by the letter O or the letter Q in most programmable controller systems. If any rung path has logic continuity, the referenced output is energized or set TRUE (logic bit = 1). The output bit is turned OFF if logic continuity is lost to the output coil.

When the output is ON, a normally opened instruction of the same address will close, and a normally closed instruction will open. In Figure 6-4, the output instruction O:0/01 is energized or TRUE if either input A or input B is TRUE or if both inputs are TRUE.

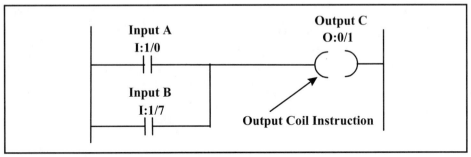

Figure 6-4. "OR" LAD program using an output instruction.

The sample applications shown in Examples 6-1 and 6-2 will help to illustrate ladder diagram programming.

EXAMPLE 6-1

Problem: Write a ladder diagram program to start and stop a pump. In this application, the normally open (NO) contacts of a local panel-mounted start pushbutton are wired to input bit address I:1/1, and the normally closed (NC) contacts of a stop pushbutton are connected to input bit address I:1/0. The pump starter relay is connected to PLC output O:3/1, and the auxiliary (aux.) motor start contacts (NO) are connected to PLC input I:1/2.

Solution: The application can be implemented using the LAD program shown in Figure 6-5.

When the NO start pushbutton (PB) is depressed, input I:1/1 is TRUE. Since the NC stop pushbutton is not depressed, input I:1/0 is also TRUE. As a result, there is logic continuity in rung 0, and the output bit O:3/1 is energized or set to one (1). Output O:3/1 energizes the pump start relay, causing the pump auxiliary (aux.) contacts to close. This, in turn, sets input bit I:1/2 to seal in the start pushbutton (PB) bit and hold the pump ON until the stop PB is depressed. After the stop PB is depressed, the stop bit I:1/0 is set to zero and the run pump output bit is deenergized or set to zero. As a result, the pump will be turned OFF, and the auxiliary contacts on the pump starter will open and set the input bit I:1/2 to zero.

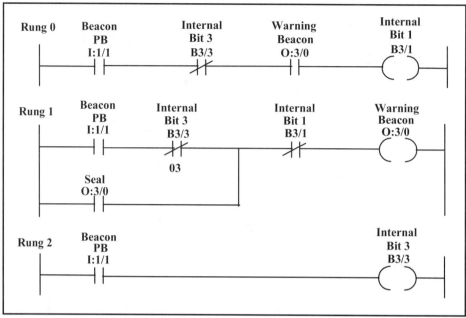

Figure 6-5. Solution for Example 6-1, pump start/stop LAD program.

It is important to note that the normally closed (NC) contacts of stop pushbuttons are always used to ensure that moving equipment applications operate safely. The NC contacts are used in the stop circuit, so if a wire from the stop PB to the PLC is cut or removed, the moving equipment will stop and it cannot be restarted. On the other hand, suppose the normally open contacts of the stop PB were used and the control wires from the PB to the PLC were cut or removed when the moving equipment was energized. In this case, depressing the stop pushbutton would not stop the equipment.

EXAMPLE 6-2

Problem: There is only a single field-mounted pushbutton available to turn an electric warning beacon on and off. Write a ladder program to control the beacon. Assume that the normally open contacts of the pushbutton are wired to input point I:1/1 and the beacon is connected to output O:3/0.

Solution: The warning beacon can be controlled using the LAD program shown in Figure 6-6.

In the sample program shown in Figure 6-6, when the beacon control pushbutton (PB) is depressed for the first time, output bit (O:3/0) is energized, and it turns ON the warning beacon. This output control bit (O:3/0) also seals itself in. If the pushbutton is depressed again, the

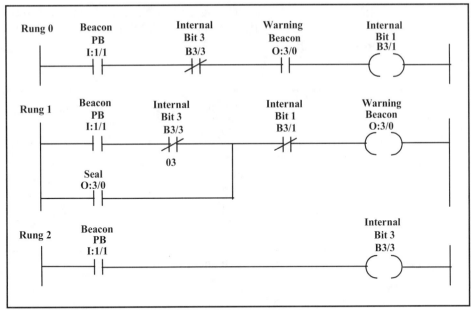

Figure 6-6. Alarm beacon LAD program.

beacon is turned OFF. The second rung (rung number 1) of the program detects the first time the pushbutton is depressed, while the first rung (rung number 0) senses the second time the pushbutton is depressed. The last rung (rung 2) is used to control internal bit 3 (B3/03). The normally closed contacts of internal bit 3 are used in rungs 0 and 1 to help perform the "push to start " and "push to stop" functions in the ladder diagram program.

Latch Coil

The latch coil instruction is programmed, if it is necessary, to ensure that an output remains energized even though the status of the input bits that caused the output to energize may change. If any rung path has logic continuity, the output is turned ON and retained ON even if logic continuity or system power is lost. The latched output will remain latched ON until it is unlatched by an output instruction of the same reference address. The unlatch instruction is the only automatic (programmed) way to reset the latched output. Although most controllers allow internal or external outputs to be latched, some are restricted to latching internal outputs only.

Unlatch Coil

The unlatch coil instruction is programmed to reset a latched output at the same reference address. If any rung path has logic continuity, the referenced address is turned OFF. The unlatch output is the only automatic way to reset a latched output, other than clearing the program. Figure 6-7 illustrates how the latch and unlatch coils are used to start or stop a process batch.

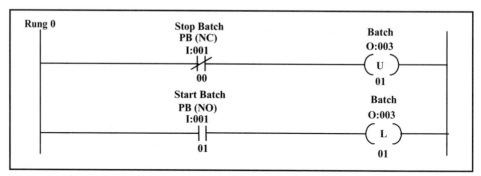

Figure 6-7. LAD program using latch and unlatch instructions.

In this example, the NO contacts of the start PB and the NC contacts of the stop PB are connected to discrete PLC inputs. The NC contacts of the stop PB are connected to input bit I:001/00, and the NO contacts of the start PB are connected to input I:001/01. If the start PB is depressed, output O:003/01 is latched ON. When the start bit (I:001/01) goes FALSE, the output instruction for the pump starter will remain ON until the stop bit (I:001/00) is depressed to unlatch the output. Note that the output unlatch instruction has the same address as the latched bit.

The LAD program in Figure 6-7 is a simpler method for producing a start/stop function then the standard start/stop LAD program shown in Figure 6-5.

One Shot (ONS)

The one shot (ONS) is an input instruction that goes TRUE for one PLC program scan if there is a FALSE-to-TRUE transition in the conditions preceding it in the rung. The one shot instruction is generally used to start operations that are triggered by momentary pushbutton action, such as when the PLC obtains values from thumbwheel switches or freezes rapidly displayed LED data. In the Allen-Bradley series 5 PLCs, the bit address must be either a binary file (B3) or an integer file (N7) bit address. A typical example is shown in Figure 6-8. In this application, when the data in pushbutton (PB) is depressed, it sets the input bit (I:001/02) to 1, and the ONS bit (B3/04) conditions the rung so the output (B3/05) turns ON for one scan. The output turns OFF for successive scans until the input goes from FALSE to TRUE again.

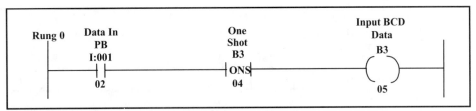

Figure 6-8. One shot (ONS) application.

Timer and Counter Instructions

Timers and counters are output instructions that provide the same functions as hardware timers and counters. They are used to activate or deactivate a device after an expired interval or count. A typical application for a counter is to count the number of parts produced on an assembly line. A typical application for a timer is to delay an operation for a fixed interval. For example, the starting of a pump might be delayed for several seconds until a valve on the discharge line of the pump is completely opened.

Timers and counters operate in quite similar ways since timers are essentially counters. Timers count the number of times that a fixed interval of time elapses. For example, to time an interval of three seconds, a timer might count three one-second intervals. A counter simply counts the occurrence of an event.

Timer Memory Word Structure

Most timer instructions require three memory registers or words: a control word or register, an accumulator (ACC) word to store the elapsed time, and a preset (PR) memory word to store a preset timer value. The preset value will determine the number of time-based intervals that are to be counted. When the accumulated value equals the preset value, a status bit is set ON and can be used to turn on an output bit.

Figure 6-9 shows a typical example of timer memory word structure for an Allen-Bradley PLC5 programmable controller system. The left-most three bits (bits 13, 14, and 15) in the timer control word are used as status bits. Bit 15 is the timer enable (EN) bit, and it is set when the logic to the timer is logical 1 or TRUE. Bit 14 is the timer timing bit (TT), and it is set when the timer rung goes TRUE. It indicates that a timing operation is in progress. Bit 13 is the timer done (DN) bit, and it is TRUE when the accumulated value is equal to the preset value.

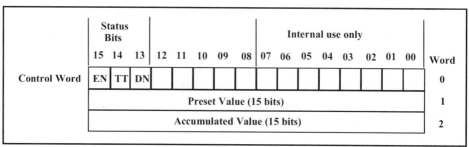

Figure 6-9. A-B PLC5 timer word structure.

Timer On Delay (TON)

The timer on delay (TON) output instruction is programmed to provide time-delayed action or to measure the duration for which some event is occurring. If any rung path connected to the input side of the timer has logic continuity, as shown in Figure 6-10, the timer begins counting time-based intervals. It counts until the accumulated (ACCUM) time equals the preset value as long as the rung conditions remain TRUE. When the accumulated time equals the preset time, a timer done (DN) bit in the timer word is set to one (1). Whenever the rung logic conditions for the TON instruction go FALSE, the accumulated value is reset to all zeros.

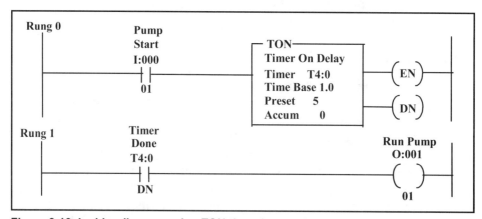

Figure 6-10. Ladder diagram using TON timer instruction.

In the sample application shown in Figure 6-10, when the pump start switch is ON, bit I:000/01 is set to one (1), and the timer (T4:0) begins to count time-based intervals. As long as the switch remains CLOSED or ON, the timer increments its accumulated value word for each time interval. When the accumulated value equals the preset value of five seconds, the timer stops incrementing its accumulated value and sets the timer done (DN) bit to ON. This done bit (T4:0/DN) is then used in rung 1 to energize the run pump output bit (O:001/01).

Timer Off Delay (TOF)

The timer off delay (TOF) output instruction provides another form of timer action. If logic continuity is lost, the timer begins counting time-based intervals until the accumulated time equals the programmed preset value. When the accumulated time equals the preset time, the timing is complete, and the timer done bit (bit 13) is reset to zero. The timer done (DN) bit can be used throughout the program as an NO or NC instruction. If logic continuity is gained before the timer is timed out, the accumulator word is set to zero and the done bit is set to a logic 1.

Figure 6-11 shows a sample program for a TOF instruction with a preset value of five seconds. In rung 0, when input I:000/01 is TRUE, the DN bit is set to one (1), turning ON the output bit, O:001/001. If the input switch (I:000/01) is OPEN for five seconds or more the timer will count up to five. When the preset value is equal to the accumulated value of five (5), the timer done bit (T4:1/DN) will be set to zero and the output bit O:001/01 in rung 1 will be turned OFF.

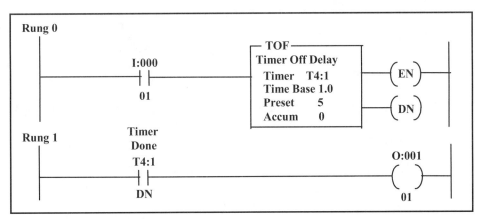

Figure 6-11. Ladder diagram using a TOF timer instruction.

Retentive Timer On (RTO)

Use the retentive timer on (RTO) output instruction if you need to retain the timer accumulated value, even if logic continuity or power is lost. If the timer rung path has logic continuity, the timer begins counting time-based intervals until the accumulated time equals the preset value. The accumulator register retains the accumulated value even if logic continuity is lost before the timer is timed out or if power is lost, as shown in Figure 6-12. When the accumulated time equals the preset time, the done bit is set. The timer done bit can be used throughout the rest of the program as NO or NC instructions. The retentive timer accumulator value is reset to zero by a reset (RES) instruction. The RES instruction

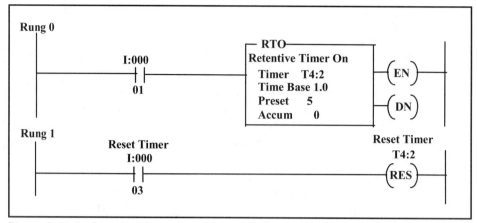

Figure 6-12. Ladder diagram using a retentive timer on instruction.

(T4:2/RES) in rung 1 will set to logical 1 if the input bit I:000/03 is set to one (1). This will reset the accumulator in timer T4:2 to zero and reset the DN bit.

A typical application for timers is to produce alternating pulses for a flashing alarm light, as illustrated in Example 6-3.

EXAMPLE 6-3

Problem: Design a timing circuit that can be used to generate an alternating signal so an alarm light connected to output at bit location O:3/1 flashes. Select a time period of 0.5 seconds OFF and 0.5 seconds ON.

Solution: The ladder logic program consists of three logic rungs with 0.5-second timers in rungs 1 and 2 and a timer done bit that drives the alarm light in rung 3, as shown in Figure 6-13.

Up Counter (CTU)

The accumulated value in the up counter (CTU) output instruction will increment by one each time there is a 0 to 1 transition of the input logic. A typical control application for a counter is to turn a device ON or OFF after a certain count is reached. Since up counters increment their accumulated value only when the up counter logic input makes a 0 to 1 transition, the rung condition must go from TRUE to FALSE and then back to TRUE before the next count is registered.

When the accumulated (ACC) value reaches the preset (PR) value, the count done (DN) bit is set to one (1). Unlike a timer instruction, the counter instruction continues to increment its accumulated value after the

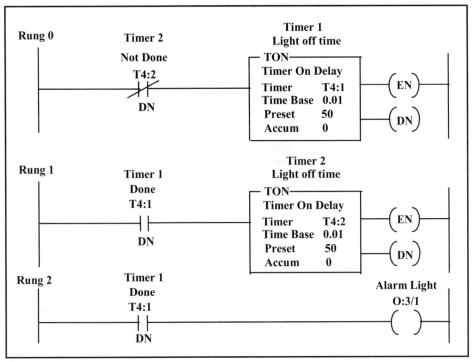

Figure 6-13. LAD program solution to problem in Example 6-3.

preset value has been reached. If the accumulated value goes above the maximum range of the counter, an overflow (OV) bit will be set. This overflow bit can be used to cascade counters for counter applications that are greater than the maximum value of the counter.

Down Counter (CTD)

The down counter (CTD) output instruction will count down by one each time a FALSE to TRUE (0 to 1) transition occurs in the input logic to the counter. In some applications, a down counter is used in conjunction with an up counter to form an up/down counter.

Figure 6-14 shows a typical example of an up/down counter using A-B PLC5 up-and-down counter instructions. Note that the same word address, C5:0, is used for both counters.

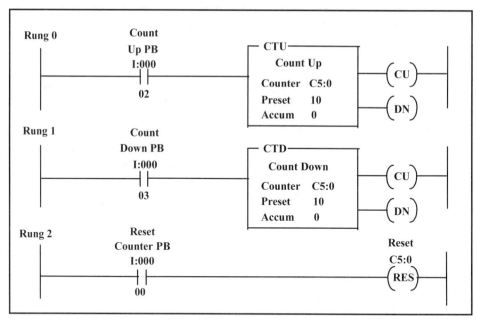

Figure 6-14. Ladder diagram for up/down counter.

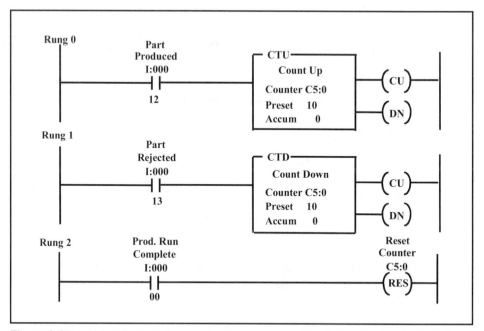

Figure 6-15. Ladder diagram for production part count (solution for Example 6-4).

> **EXAMPLE 6-4**
>
> **Problem:** Write an LAD program using counter instructions to count the number of parts produced on an assembly line. Assume that input I:000/12 of the PLC is activated by each part leaving the final assembly line, that input I:000/13 is activated by a part being rejected in final test, and that input I:000/00 is energized at the end of a production run.
>
> **Solution:** Figure 6-15 gives the ladder program for calculating the number of parts produced. The number of parts produced is the value found in the accumulator of the counter C5:0.

Data Transfer Operations

Data transfer instructions involve transferring the contents of one register to another. Data transfer instructions can address any location in the memory data table, with the exception of areas restricted to user applications. Prestored values can be automatically retrieved and placed in any new location. That location may be the preset register for a timer or counter or it may even be an output register that controls a seven-segment display.

The Allen-Bradley PLC5 programming system uses three data bit and word transfer instructions: bit distribute (BTD), move (MOV), and masked move (MVM). These data transfer instructions are used by most PLC manufacturers.

Bit Distribute (BTD)

The BTD instruction is an output instruction that moves up to sixteen bits of data within or between words. The source of the data remains unchanged. Figure 6-16 shows a BTD example of moving bits within a word. The instruction writes over the destination with the specified bits. If the length of the bit field extends beyond the destination word, the processor does not save the overflow bits. Because they do not wrap into the next word they are lost.

On each processor program scan, when the rung that contains the BTD instruction is TRUE, the processor moves the bit field from the source word to the destination word. To move the data within a word, the programmer selects the same word address for both the source and the destination, as shown in Figure 6-16. In this example, four bits are moved from the left-hand side (bits 00 through 03) of word N70:00 to the middle of the word (bits 08 through 11).

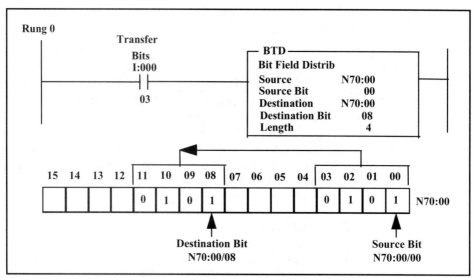

Figure 6-16. Example of BTD instruction moving bits within a word.

MOVE (MOV) and MASKED MOVE (MVM)

The move (MOV) instruction is an output instruction that copies the data in a source address to a destination address. As long as the rung remains TRUE, the instruction moves the contents of each scan of the PLC processor to the destination address. The source can be a program constant or data address from which the instruction reads an image of the value. The destination is the data address to which the instruction writes the result of the operation. The instruction writes over any data stored at the destination. For example, in rung 0 of Figure 6-17, when input I:000/02 is TRUE, the data stored at address N7:10 is copied and written into location N7:12.

The masked move (MVM) instruction is an output instruction that copies the source to a destination and allows portions of the data to be masked. As long as the rung remains TRUE, the instruction moves data each scan.

The MVM instruction can be used to copy I/O image table, binary, or integer values. For example, it can be used to extract bit data, such as status or control bits, from an address that contains bit and word data. The source is a program constant or data address from which the instruction reads an image of the value. The source remains unchanged.

The mask can be an address or hexadecimal value that specifies which bits to pass or block. The programmer must set the mask bits to one (1) to allow the data to pass to the destination. The moved data overwrites the destination data. The destination is the data address to which the instruction writes the result of the operation. The instruction writes over any data stored at the destination.

Figure 6-17 shows an example of both move and masked move instructions. In rung 0, when the input bit I:000/02 is TRUE the contents of memory location N7:10 are moved to integer word location N7:12. In rung 1, when the input bit I:000/02 is TRUE the contents of the higher eight bits of memory word N7:10 are moved to integer word location N7:12 since the higher eight bits of the mask (FF00) contain all ones (two hex "F") . The lower eight bits are blocked because the lower eight bits of the mask are all zeros.

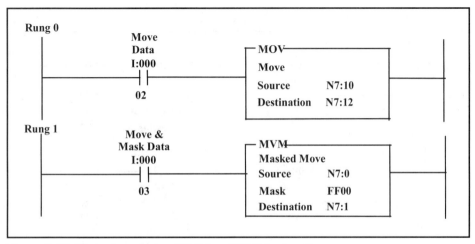

Figure 6-17. Typical example of MOV and MVM instructions.

Arithmetic Operations

The arithmetic operations include the four basic operations of addition, subtraction, multiplication, and division. These instructions use the contents of two word locations to perform the desired function.

The arithmetic instructions are output instructions that may or may not have input logic instructions. They use either one or two data words to store the result. The add and subtract instructions use one word. Multiply and divide need two words for the computed result. We will discuss each instruction in turn in the following sections.

Addition (ADD)

The addition (ADD) instruction performs the addition of two values stored in two different memory locations, source A and source B. The result is stored in the destination register. Source A and source B can be either values or addresses that contain values. In the example shown in Figure 6-18, if the input I:000/01 is TRUE, the instruction adds the value stored in N7:0 to the value stored in N7:1 and stores the result in address N7:2.

Subtraction (SUB)

The subtraction (SUB) instruction performs the subtraction operation of two numbers located in source A and B. As in addition, if a condition to enable the subtraction is set to one (1), the result is placed in a destination. Looking at rung 1 in Figure 6-18, if the input bit at address I:000/02 is set to one (1), the number in address N7:4 is subtracted from the number in address N7:3 and the result is placed in register N7:5.

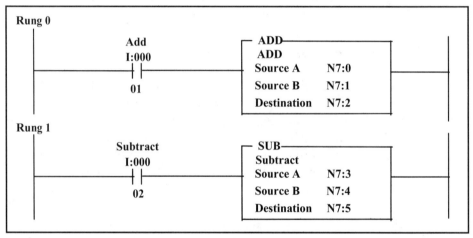

Figure 6-18. Typical example of ADD and SUB instructions.

Multiplication (MUL)

The multiplication (MUL) instruction is used to multiply one value (source A) by another value (source B) and to place the result in the destination. Source A and source B can be values or addresses. In the typical MUL instruction shown in Figure 6-19, if input bit I:000/03 in rung 0 is TRUE, the processor will multiply the value in N7:3 by the value in N7:4 and store the result in N7:20.

Division (DIV)

The division (DIV) instruction is used to divide one value (source A) by another value (source B) and to place the result in the destination. Source A and source B can be values or addresses. In the typical DIV instruction example shown in Figure 6-19, if input bit I:000/02 in rung 1 is TRUE, the processor will divide the value in N7:0 by the value in N7:1 and store the result in N7:21.

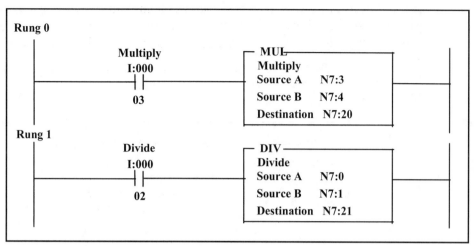

Figure 6-19. Typical examples of MUL and DIV instructions.

Data Comparison Operations

In general, manipulating data using ladder diagram instructions involves using simple register (word) operations to compare the contents of two registers. In the ladder diagram language, there are three basic data comparison instructions: equal to, less than, and greater than. Depending on the result of a greater-than, less-than, or equal-to comparison, an output can be turned ON or OFF or some other operation can be performed.

Equal To (EQU)

The equal to (EQU) instruction is used to test whether two values are equal. Source A and source B can be either values or addresses that contain values. For example, in rung 0 of Figure 6-20, if the equality is TRUE the output coil is energized.

Less Than (LES)

Much like the equal to instruction, the less than (LES) instruction tests the contents of the value of one location (source A) to see if it is less than the value stored in a second location (source B). If the test condition is TRUE, the output coil in rung 1 in Figure 6-20 is energized.

Greater Than (GRT)

The greater than (GRT) instruction operates like the less than operation, with the exception that the test is performed for a greater than condition. If the test condition is TRUE, the output coil in rung 2 of Figure 6-20 is energized.

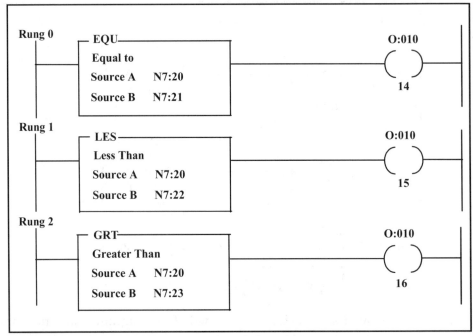

Figure 6-20. Examples of data comparison instructions.

Program Control Instructions

The program control functions are used to perform a series of conditional and unconditional jump and return instructions. These instructions allow the program to execute only certain sections of the control logic if a fixed set of logic conditions is met. The instructions described in the following sections are a representative selection of some of the program control instructions available in most controllers.

Master Control Relay (MCR)

The master control relay (MCR) instruction is used in pairs to activate or deactivate the execution of a group or zone of ladder rungs. A conditional MCR instruction is used in conjunction with another unconditional MCR coil to place a fence around the group of rungs. For example, in Figure 6-21, if the input I:000/03 is TRUE, the conditional MCR coil in rung 0 will be energized and the logic inside the zone will be executed according to the logic in each rung inside the MCR zone. If the conditional MCR instruction is turned OFF, all nonretentive outputs inside the zone will be deenergized.

The rungs within an MCR zone are still scanned, but the PLC scan time is reduced because of the false state of the nonretentive outputs. Nonretentive outputs are reset when their rung goes FALSE.

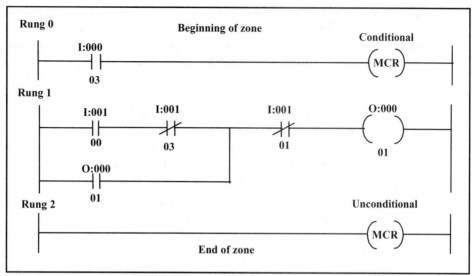

Figure 6-21. A typical LAD program using MCR instructions.

Jump (JMP) and Label (LBL)

The jump (JMP) and label (LBL) instructions are used in pairs to skip portions of the ladder logic program. The jump instruction allows the normal sequential execution to be altered so the CPU will jump to a new position in the ladder program. If the jump rung logic is TRUE, the jump coil (JMP) instructs the CPU to jump to and execute the rung that is labeled with the same reference address as the jump coil. This allows the program to execute rungs that are out of the normal sequential flow of a standard ladder program.

The purpose of the label is to identify the ladder rung that is the destination of a jump instruction. The label reference must match that of the jump instruction with which it is used. The label instruction does not contribute to logic continuity, and it is always logically TRUE. It is placed as the first logic condition in the rung. A label instruction referenced by a unique address can be defined only once in a program.

Jump to Subroutine (JSR), Subroutine (SBR), and Return (RET)

Subroutines are used in programming to produce a more structured program and to reduce the amount of memory used for a program. Subroutines are used to store recurring logic functions that can be accessed from different parts of the main ladder logic program. This saves memory space because the function has to be programmed only once even though it is used many times in the control application.

The Allen-Bradley (A-B) PLC5 has three subroutine instructions: the jump to subroutine (JSR), the subroutine (SBR), and the return (RET). These instructions direct the processor to go to a separate subroutine file within the ladder logic program, scan that subroutine file once, and return to the point of departure.

The JSR instruction directs the processor to the specified subroutine file and, if required, defines the data passed to and received from the subroutine. The optional SBR instruction is the instruction that stores incoming data. The SBR instruction is only used if the ladder diagram requires that data be passed to and from the subroutine. The RET instruction ends the subroutine and, if required, stores data to be returned to the JSR instruction in the main program. If the SBR instruction is used, it must be the first instruction on the first rung in the program file that contains the subroutine.

PLC Control Program Documentation

An important part of PLC programming is documenting the control program properly and completely. Most manufacturers' PLCs make it possible to print out a hard copy of the control program stored in the PLC's memory. Whether stored in ladder diagram form or in some other language, the hard copy will be an exact replica of the control program stored in memory.

This hard-copy printout will show each programmed instruction with the associated address of each input and output. However, the information indicating the function or purpose of each field device or internal control bit or instruction is not readily apparent. Additional documentation is generally required. Most PLC manufacturers provide software documentation programs that allow the programming device, generally a personal computer, to enter labels or mnemonic nomenclature for each program element or instruction. The examples of ladder diagram programs provided in this chapter illustrate documented elements or instructions in the ladder rungs such as input, output, timer, counter, and so on. Most PLC programming software packages also allow general comments to be added on each rung of logic, as shown in Figure 6-22.

The PLC controller will always store the latest revision of the control program in memory. So before testing a program on line, the user should print out the latest revision of the program. During start-up and testing, frequent changes are made to the control program. The user should document these immediately with both rung and instruction comments. It is a good practice to obtain the latest hard copy of the PLC program before performing maintenance on a system.

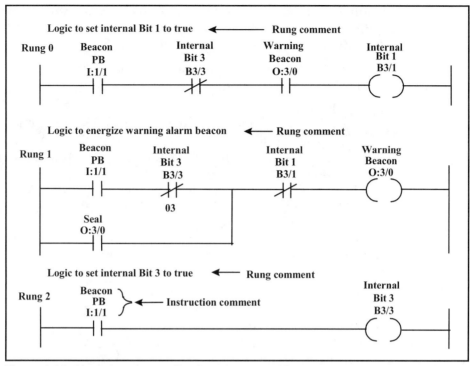

Figure 6-22. Alarm beacon application program with rung comments.

EXERCISES

6.1   Write a ladder logic diagram program to start a pump that fills a process tank with fluid until a high level is reached in the tank. Assume that the fill tank pushbutton (PB1) is wired to an Allen-Bradley PLC5 discrete input module at bit I:010/00 and that a NC tank high-level switch is connected to input bit I:010/01. Also assume that the pump start relay is connected to output O:000/01 on the PLC.

6.2   Write a ladder diagram program to control an electric motor. Assume that a normally opened start pushbutton is wired to an input module at I:000/00 and that a normally closed stop pushbutton is connected to the same input module at address I:000/01. Also, assume that PLC5 output O:001/03 is connected to the motor starter and that the normally opened auxiliary contacts on motor start contactor are connected to PLC input I:000/02.

6.3   Write a ladder logic program to control a process pump under the following conditions: the pump is turned on five seconds after both the inlet and outlet valves to the pump have been opened and the pump is turned off, if either the inlet or the outlet valve are closed. Assume the following: a pump starter is connected to output bit O:001/00, a valve-open position switch on the inlet valve is wired to input I:000/02, and a valve-open position switch on the outlet valve is wired to input I:000/03.

6.4   Write a LAD program to open the control valve (LV-1) on the outlet line of the process tank shown in Figure 6-23 if the following conditions are met: the normally opened level switch (LSH-1) closes when the automatic position on the HOA (HS-1) is selected or the operator places the HOA in the hand position. Assume that the high-level switch is connected to input I:002/01 and that the output valve is connected to output O:003/01 on an Allen-Bradley PLC5. Also, assume that the "Auto" position of the HOA switch is connected to input I:002/02 and that the "Hand" position of the switch is wired to input I:002/03 on the PLC5.

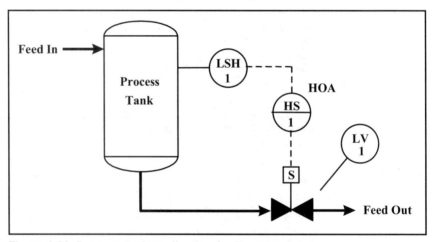

Figure 6-23. Process tank application for Exercise 6.4.

6.5   Write a LAD program using timer instructions to flash two process alarm lights on a control panel. The two alarm lights are driven by PLC output signals O:010/00 (alarm 1) and O:010/01 (alarm 2). Assume that alarm 1 is activated by input I:000/01 and that alarm 2 is activated by input I:000/02.

6.6   Write a ladder diagram program to control the temperature of the fluid in a process tank so it is close to 400°C. Assume that the heater contactor is connected to PLC5 output O:001/01, a temperature-low switch set at 395°C is connected to input I:000/00, and a temperature-high switch set at 405°C is connected to input I:000/01. The temperature-low switch is closed if the fluid temperature is below 395°C and is opened at a temperature of 395°C or higher. The temperature-high switch is closed if the fluid temperature is below 405°C and opened at 405°C or higher temperatures.

6.7　Write a LAD program for an Allen-Bradley PLC5 that turns off a conveyor belt on a production line after fifty parts have been produced. Assume the following: (1) when output bit O:001/12 is set to 1 the conveyor belt is turned on, (2) input I:002/01 changes from 0 to 1 and then back to 0 each time a production part is rejected, (3) input I:002/02 changes from 0 to 1 and then back to 0 each time a new part is produced, (4) a normally open (NO) pushbutton connected to input I:002/03 is used to set the production count to fifty, and (5) a NO pushbutton connected to input I:002/04 is used to reset the counter to zero and to stop the conveyor belt.

6.8　Write an A-B PLC5 program to add the integer number in word N7:2 to the integer number in word N7:4. Then divide the result stored in N7:20 by five and store the final result in word N7:6.

6.9　Write a ladder logic program to transfer the integer number in word N7:0 to an output display at word N7:10 if the number in word N7:0 is greater than the number in word N7:1 and less than the number in location N7:2.

6.10　Explain the purpose and benefits of using subroutines in PLC programming applications.

BIBLIOGRAPHY

1.　Allen-Bradley Co., Inc. *Programmable Controller Fundamentals* (Allen-Bradley, 1985).

2.　Bryan, E. A., and L. A. Bryan. *Programmable Controllers: Concepts and Applications* (Industrial Text Co., 1988).

3.　Gilbert, R. A., and J. A. Llewellyn. *Programmable Controllers: Practices and Concepts* (Industrial Training Corp., 1985).

4.　Jones, C. T., and L. A. Bryan. *Programmable Controllers: Concepts and Applications* (International Programmable Controls, 1983).

5.　Reis, R. A., and J. W. Webb. *Programmable Controllers: Principles and Applications*, 3d ed. (Prentice-Hall, 1995).

6.　Rockwell Software, Inc. *PLC-5 Programmable Controller: Instruction Set Reference* (Rockwell Software, 1996).

7

Advanced LAD Programming

Introduction

The advanced ladder diagram instructions are required to perform more powerful functions beyond simple ON/OFF control, timing, counting, and data manipulation. These advanced instructions are used for analog control, data file operations, sequencer operations, data reporting, complex logic functions, and other functions that are not possible with the basic LAD instructions.

Advanced LAD Instructions

Advanced LAD instructions allow the user to program more complex PLC control functions. In this chapter, we will discuss the most common advanced instructions-file, shift register, sequence, and block transfer-for the Allen-Bradley PLC5 family of PLCs.

File Instructions

A file is a group of consecutive data table words that are used to store PLC information. A file instruction is used to perform arithmetic, logical, search, copy, and compare operations. We will discuss the Allen-Bradley PLC5 file instructions—file arithmetic and logical (FAL), file search and compare (FSC), file copy (COP), and file fill (FLL).

Figure 7-1 illustrates the structure of a typical FAL file instruction with its parameters of control, length, position, mode, destination, and expression.

The processor uses this information to run the instruction. The *control* is the address of the control structure in a control type (R) file. The *length* is

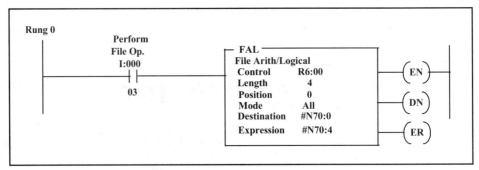

Figure 7-1. Typical A-B PLC5 file operation instruction.

the number of words (0 to 999) in the data block on which the file instruction operates. The *position* is the current element within the data block that the processor is accessing. The *mode* is the number of file elements operated on each time the rung is scanned in the program. There are three modes: All, Numerical, and Incremental. In the All mode, the entire file is operated on before the processor continues on to the next rung of the program. The Numerical mode distributes the file operation over a number of program scans. The Incremental mode manipulates one word of the file each time the rung goes from FALSE to TRUE. The *destination* is the address where the processor stores the result of the operation. The instruction converts to the data type specified by the destination address. The *expression* contains addresses, program constants, and operators that specify the source of data and the operations to be performed.

The output coils to the right of the file instruction are the enable (EN), done (DN), error (ER) bits. These bits have the same word address as the instruction control. The processor automatically sets the address of these status bits when the programmer enters the control address. The enable (EN) bit is set by a FALSE-to-TRUE rung transition, and it indicates that the instruction is enabled. In the Incremental mode, the EN bit follows the rung condition. In the Numerical and All modes, the EN bit remains set until the instruction completes its operation, regardless of the rung condition. The enable bit is reset when the rung goes FALSE, and the instruction then completes its operation.

The done (DN) bit is set after the instruction has operated on the last group of words. In the Numerical mode, if the instruction is FALSE at completion, the instruction has the processor reset the done bit one program scan after the operation is complete. If the instruction is TRUE at completion, the done bit is reset when the instruction goes FALSE.

The error (ER) bit is set when the operation generates an overflow. The instruction stops until the ladder program resets the error bit. When the processor detects an error, the position value stores the number of the word that faulted.

File Arithmetic and Logic (FAL)

The file arithmetic and logic (FAL) instruction performs copy, arithmetic, logic, and function operations on the data stored in files. The FAL instruction is an output instruction that performs the operations defined by the source addresses and the operators listed by the programmer in the expression field. The instruction writes the results into a destination address. The FAL instruction automatically converts the data type at the source addresses into the data type that is specified in the destination address. The FAL instruction performs operations such as zeroing a file, copying data from one file to another, making arithmetic or logic computations on data stored in files, and unloading a file of error codes one at a time for display. Table 7-1 lists the operations performed by the A-B PLC5 FAL instruction.

Table 7-1. FAL Operators for A-B PLC5 Family

Type	Operator	Description	Example
Copy	none	Copy from A to B	
Clear	none	Set a value to 0	
Arithmetic	+	Add	2+2
	-	Subtract	8-5
	*	Multiply	3*6
	/	Divide	12/4
	-	Negate	-N7:0
	SQR	Square root	SQR N7:1
	**	Exponential	10**2
Bitwise	AND	Bitwise AND	D9:3 AND D10:4
	OR	Bitwise OR	D9:4 OR D9:5
	XOR	Bitwise exclusive or	D9:6 XOR D9:7
	NOT	Bitwise complement	NOT D10:11
Conversion	FRD	Convert from BCD to binary	FRD D10:0
	TOD	Convert from binary to BCD	TOD N7:1

To illustrate how a FAL instruction operates, we will perform a FAL copy operation, as shown in Figure 7-2.

In this example, when the rung goes TRUE (bit I:000/02 set to 1), the processor reads the data stored in four words of integer file N71, starting at word 3. It then writes the data to integer file N70 starting at word 0. It writes over any data in the destination file.

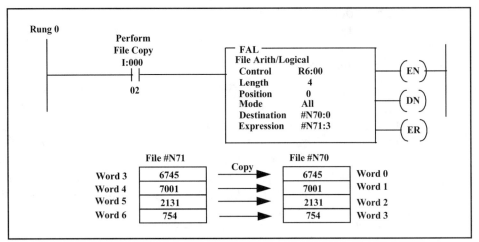

Figure 7-2. A-B PLC5 file-to-file copy example.

To illustrate a FAL instruction, a sample program is shown in Example 7-1.

EXAMPLE 7-1

Problem: Write a PLC5 LAD program to copy the data in integer file N30, words 5, 6, and 7, to file N31 starting at word 2, if input bit I:000/03 is TRUE.

Solution: The ladder diagram program to copy the data is shown in Figure 7-3.

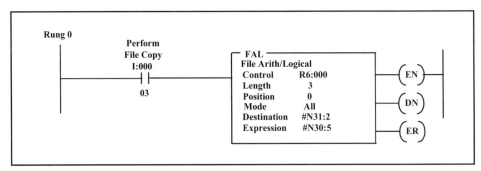

Figure 7-3. File-to-file copy LD for Example 7-1.

File Search and Compare (FSC)

The file search and compare (FSC) instruction is an output instruction that compares values in source files, word by word, for the logical operation specified in the expression. When the processor finds that the specified comparison is TRUE, it sets the found (FD) bit and records the position where the TRUE comparison was found. The inhibit (IN) bit is set to prevent any further searching of the files.

The FSC instruction is used to perform operations such as setting high and low process alarms for multiple analog inputs and comparing batch variables against a reference file before starting a batch operation. The FSC instruction performs the comparisons listed in Table 7-2 on file data according to the equation listed in the expression portion of the instruction. The processor compares files of different data types by internally converting data into its binary equivalent before performing the comparison.

Table 7-2. FSC Comparison Instructions for A-B PLC5s

Comparison	Sample Expression
Search equal	#N50:0 = #N51:0
Search not equal	#N51:0 <> #N53:10
Search less than	#N51:0 < #N53:10
Search less than or equal	#N51:0 <= #N53:10
Search greater than	#N51:0 > #N53:10
Search greater than or equal	#F61:0 >= #N63:10

In file search, when the rung condition is TRUE, the desired comparison is performed on data addressed in the expression. Words are compared in ascending order starting at the beginning. The rate is determined by the mode of operation specified in the FSC instruction. The done (DN) bit is set after the processor has compared the last pair. If the rung is TRUE at completion, the done bit is turned off when the rung is no longer TRUE. In the numerical mode, however, if the rung is not TRUE at completion, the DN bit stays on one program scan after the operation is complete.

To illustrate how a FSC instruction operates, we will perform a FSC "search not equal" operation, as shown in Figure 7-4. When bit I:000/03 goes to TRUE, the processor performs the "not-equal-to" comparison between words, starting at B3:0 and B15:0. The number of words compared per program scan is ten in this example because the mode is set to ten (10).

When the processor finds that corresponding source words are not equal (words *B3:4* and *B15:4*), it stops the search and sets the found (FD) and inhibit (IN) bits. To continue the search comparison the ladder logic program must turn off the inhibit bit.

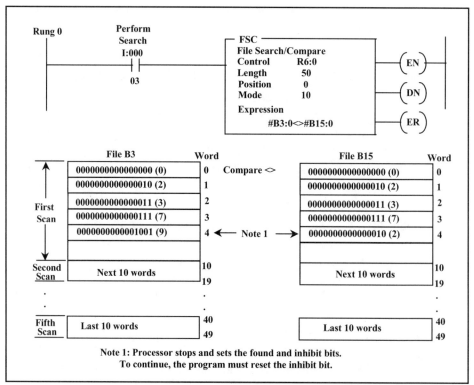

Figure 7-4. Typical A-B PLC5 FSC instruction example.

A sample ladder diagram program is shown in Example 7-2 to illustrate a FSC instruction.

EXAMPLE 7-2

Problem: Write a PLC5 LAD program to search the data in integer file N40, words 0 through 99, and to compare it for an equal condition to the data in file N50 starting at word 0, if input bit I:000/03 is TRUE.

Solution: The ladder diagram program to search and compare the data files is shown in Figure 7-5.

File Copy (COP)

The file copy (COP) instruction is an output instruction that copies the values in the source file into the destination file. The source file remains unchanged. The COP instruction does not use status bits. The COP instruction will not write over file boundaries, so any overflow data will be lost. Also, no data conversion occurs so the source and destination files should use the same data type.

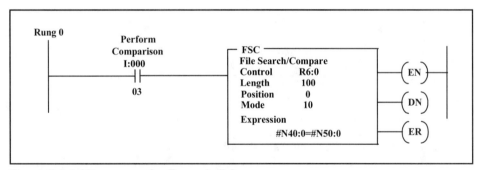

Figure 7-5. LAD program for Example 7-2.

Figure 7- 6 shows an example of a LAD program using a COP instruction. In this example, if input bit I:000/03 is TRUE, the processor will copy the first ten words starting at file N50:0 into the first ten words of file N60:0.

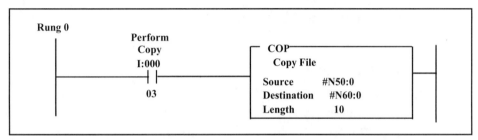

Figure 7-6. PLC5 file copy instruction.

File Fill (FLL)

The file fill (FLL) instruction is an output instruction that fills the words of a file with a source value. The source file remains unchanged. Like the COP instruction, the FLL instruction does not use status bits. The FLL instruction will not write over file boundaries, so any overflow data will be lost. Also, no data conversion occurs, so the source and destination files should use the same data type.

Figure 7-7 shows an example of a LAD program that uses a FLL instruction. In this example, if input bit I:000/03 is TRUE, the processor will copy the first ten words starting at file N50:0 into the first ten words of file N60:0.

Shift Register Instructions

Shift register instructions are used to track the movement or flow of parts and information in industrial applications. We will discuss four A-B PLC5 shift register instructions that are in common use: bit shift left (BSL), bit shift right (BSR), first-in-first out load (FFL), first-in-first out unload (FFU).

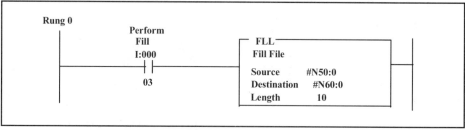

Figure 7-7. A-B PLC5 file fill instruction.

The bit shift instructions, BSR and BSL, are used to load bits into, shift bits through, and unload bits from a bit array one bit at a time. A typical application would be tracking bottles through a bottling line, where each bit represents a bottle. The load and unload shift instructions, FFL and FFU, are used to load and unload values in the same order. A typical application might be tracking parts through an assembly line where the parts are represented by values that have a part number and assembly code.

Bit Shift Instructions

Bit shift instructions shift all bits within the specified address by one bit position after every FALSE-to-TRUE ladder rung transition. There are two shift instructions: bit shift left (BSL) and bit shift right (BSR). Figure 7-8 shows the structure of a bit shift left instruction with its parameters of file, control, bit address, and length.

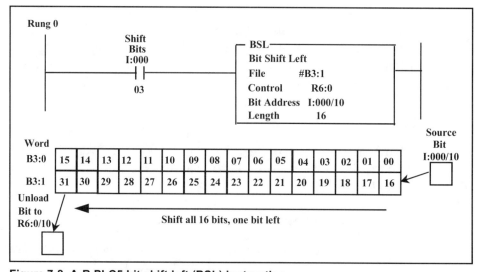

Figure 7-8. A-B PLC5 bit shift left (BSL) instruction.

The *file* is the address of the bit array that will be manipulated. The array must start at a sixteen-bit word boundary. For example, use bit 0 of word number 1, 2, 3, or so on. The array can end at any bit number up to 15,999. However, the remaining bits in the last word of the array cannot be used because the instruction invalidates them.

The *control* is the address of the control structure (forty-eight bits or three sixteen-bit words) in a control type (R) file that stores the instruction's status bits, the size of the array (number of bits), and the bit pointer. The processor uses this information to run the instruction. The *bit address* is the address of the source bit. The bit shift instruction inserts the value (0 or 1) of this bit in either the first (lowest) bit position (for the BSL instruction) or the last (highest) bit position (for the BSR instruction) in the array.

The *length* is the decimal number of bits to be shifted. In A-B PLC5s, the bits in I/O files are numbered in octal 00 to 07 and 10 to 17, but all other files are numbered in decimal 0 to 15.

The output coils to the right of the file instruction are the enable (EN) and done (DN) bits. These bits have the same word address as the instruction control element. The address of these status bits is automatically set by the processor when the programmer enters the control address. The enable (EN) bit is set by a FALSE-to-TRUE rung transition and indicates that the instruction is enabled. The done (DN) bit is set to indicate that the bit array shifted one bit position.

The control element also contains two other status bits: error (ER) at bit 11 and unload (UL) at bit 12. The error (ER) bit is set to indicate that the instruction detected an error, such as entering a negative file length. The unload (UL) bit is the instruction's output. The UL bit stores the status of the bit that was removed from the array each time the instruction is enabled.

In the LAD program shown in Figure 7-8, when the rung containing the BSL instruction goes from FALSE to TRUE, the processor sets the EN bit. The processor then shifts 16 bits (length = 16) in bit file B3, starting with bit 16, to the left (higher bit number) one bit position. The last bit shifts out at bit position 31 into the control register UL bit (R6:0/10). The source bit comes from I:000/10 and shifts into the first bit position of B3:1. After the processor completes the shift operation in one program scan, when the rung control bit I:000/03 goes FALSE, the instruction resets the status bits EN, ER (if set), and DN.

For wraparound operation, the source address is assigned the same address as the highest (outgoing) bit address.

Figure 7-9 shows the structure of a bit shift right instruction with its parameters of file, control, bit address, and length. The instruction works in the same way as a bit shift left instruction, except the bits move to the right. In the example in Figure 7-9, when the rung containing the BSR instruction goes from FALSE to TRUE, the processor sets the EN bit. Then the processor shifts sixteen bits (length = 16) in bit file B3 to the right (to a lower bit number) one bit position, starting with bit 31. The last bit shifts out at bit position 16 into the control register UL bit (R6:0/10). The source bit comes from I:000/10 and shifts into the highest bit position, bit 31 of bit file B3.

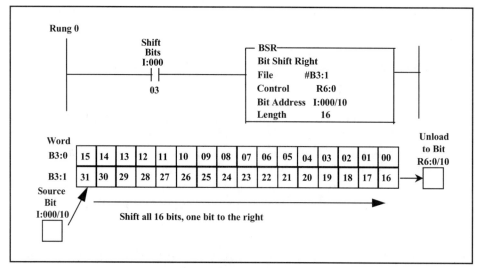

Figure 7-9. PLC5 bit shift right (BSR) instruction.

After the processor completes the shift operation in one program scan, when the rung control bit I:000/03 goes FALSE, the instruction resets the status bits EN, ER (if set), and DN. For wraparound operation, the source address is assigned the same address as the highest (outgoing) bit address.

First In–First Out (FIFO) Instructions

There are two FIFO instructions. The FIFO load uses the mnemonic FFL, and the FIFO unload uses the mnemonic FFU. These two instructions, FFL and FFU, are used in pairs to store and retrieve data in a prescribed order. When used in pairs, they establish an asynchronous shift register or stack.

The two FIFO instructions, load and unload, must use the same file and control addresses, length, and position values as shown in the example in Figure 7-10. This figure illustrates the FFL and FFU instructions with their parameters of source, destination, FIFO, control, length, and position.

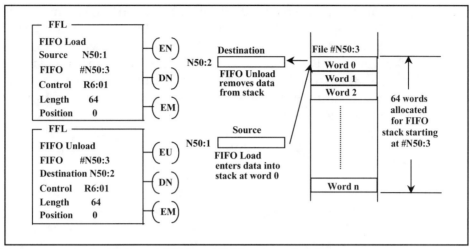

Figure 7-10. PLC5 FIFO load (FFL) and FIFO unload (FFU) instructions.

Data is loaded starting at word 0 and increasing to word n. It is unloaded from word 0, and all words are shifted up one position toward word 0.

The *Source* is the address that stores the "next in" value to the stack. The FIFO load instruction (FFL) retrieves the value from this address and loads it into the next word in the stack. The *Destination* is the address that stores the value that exits from the stack.

The *FIFO* is an indexed address of the stack. The same FIFO address is used for the associated FFL and FFU instructions. The *Control* is the address of the control structure (forty-eight bits, or three sixteen-bit words) in the control area of memory. The control structure stores the instruction's status bits, stack length, and next available position (pointer) in the stack.

The *Length* specifies the maximum number of words in the stack. The length is addressed by adding the mnemonic .LEN to the control address (R register). The *position* indicates the next available location where the instruction loads data into the stack. The position is addressed by adding the mnemonic POS to the control address (R register). Normally, the position value is 0, unless the programmer wants the instruction to start at an offset at power-up.

The output coils to the right of the FIFO instructions are the enable load (EN), enable unload (EU), empty (EM), and done (DN) bits. These bits have the same word address as the instruction control element (R). The processor automatically sets the address of these status bits when the programmer enters the control address. The enable load (EN) bit used in the FFL instruction is set by a FALSE-to-TRUE rung transition. It indicates that the instruction is enabled. The enable unload (EU) bit used in the FFU instruction is set by a FALSE-to-TRUE rung transition. It also indicates that the instruction is enabled. The done (DN) bit is set to indicate that the stack is full. The DN bit inhibits loading the stack until there is room. Empty (EM) is set by the processor to indicate that the stack is empty. The FIFO unload command should not be enabled if the EM bit is set. Example 7-3 illustrates how FIFO instructions operate.

EXAMPLE 7-3

Problem: An assembly line produces one hundred machine parts on each production run. The serial number for each part is located in word N40:1. Write a ladder logic program to place the serial numbers into File N50:3. Download serial numbers to word N40:2.

Solution: The required ladder diagram is shown in Figure 7-11.

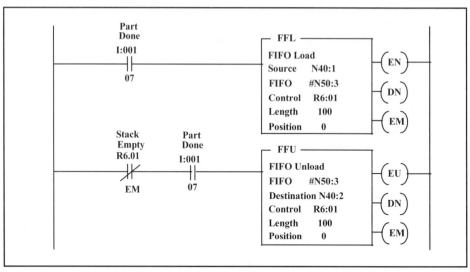

Figure 7-11. FFL and FFU ladder diagram for Example 7-3.

Sequencer Instructions

The sequencer instructions are typically used to control automatic assembly machines that have a consistent and repeatable operation. There are three common sequence instructions: sequencer input, sequencer output, and sequencer load.

The sequencer instructions are generally used to transfer data from the memory to discrete output modules for the purpose of controlling sequential process operations or sequential batch operations (sequencer output). They are also used to compare I/O word data with data stored in tables so process operating conditions can be examined for control and diagnostic purposes (sequencer input). The instruction is also used to transfer I/O word data into the memory (sequencer load).

The sequencer instructions can conserve program memory because they can monitor and control multiples of sixteen discrete outputs simultaneously in a single rung of logic. The sequencer input instruction (SQI) and the sequencer output instruction (SQO) are used in pairs to monitor and control a sequential operation, respectively.

Figure 7-12 illustrates the three A-B PLC5 sequencer instructions. The parameters used in these instructions are file, mask, source, destination, control, length, and position. The *file* is the indexed address of the sequencer file where the instruction transfers data to and from memory. Its purpose depends on the instruction.

The *mask* (for SQO and SQI) is a hexadecimal code or the address of the mask element or file through which the instruction moves data. Set mask bits to one (1) to pass data; set mask bits to zero to prevent the instruction from operating on corresponding destination bits. The programmer can specify a hexadecimal value to obtain a constant mask value. The programmer can also store the mask in an element or file if he or she wants to change the mask according to a control application requirement.

The *source* (for SQI and SQL) is the address of the input element or file from which the instruction obtains data for its sequencer file. The

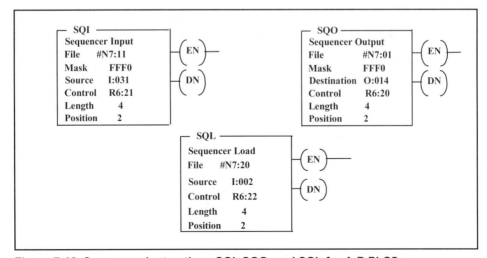

Figure 7-12. Sequencer instructions SQI, SQO, and SQL for A-B PLC5s.

destination (for SQO only) is the destination address of the output word or file to which the instruction moves data from its sequencer file. The *control* is the address of the control structure in the control area (R) of memory (forty-eight bits, or three sixteen-bit words) that stores the instruction's status bits, the length of the sequencer file, and the instantaneous position in the file.

The programmer should use the control address with the mnemonic to address the *length* (LEN) and *position* (POS) parameters. Length is the length of the sequencer file, and position is the current position of the word in the sequencer file that the processor is using. The length is the number of steps in the sequencer file starting at position 1. Position 0 is the start-up position. The instruction resets to position 1 at each completion. Note that the address assigned for a sequencer file is step 0. Sequencer instructions use (length +1) words of data for each file referenced in the instruction. This also applies to the source, mask, and destination values if they are addressed as files.

The position is the word's location in the sequencer file. Its value is incremented internally by SQO and SQL instructions. To illustrate how sequencer instructions operate, we will perform a sample sequencer output (SQO), as shown in Figure 7-13. In this example, the SQO instruction moves the data of the current step (1) through a mask to an output word that is connected to an A-B PLC5 output module in rack 1, I/O group 4.

The SQO instruction steps through the sequencer file of sixteen-bit output words whose bits have been set to control the various output devices connected to the discrete output module shown in Figure 7-13. When the rung goes from FALSE to TRUE, the instruction increments to the next step (word) in the sequencer file: #N7:1. Word *N7:1* is the "home" position or step 0, and *N7:2* is the word for step 1. When the sequence is finished it will start over again at step 1. The data in the sequencer file is transferred through a fixed mask (0F0F, HEX) to the destination address O:014. Current data is written to the destination element for every scan in which the rung remains TRUE.

Block Transfer Instructions

In larger PLC systems, large blocks of data must be transferred between supervisor PLC processors and remote processors on a high-speed data communications network. In the A-B PLC5s, block transfers are performed using block-transfer write (BTW) and block-transfer read (BTR) instructions.

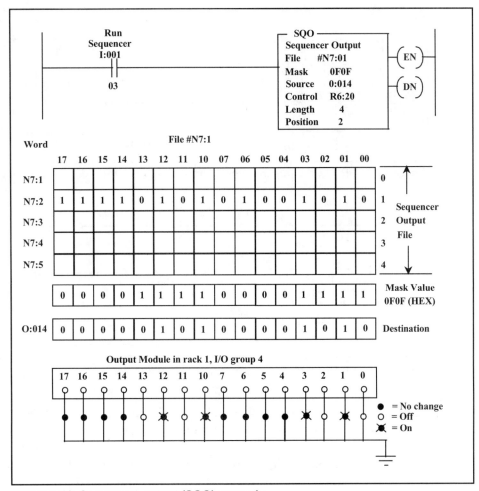

Figure 7-13. Sequencer output (SQO) example.

The basic A-B PLC5 processor in scanner mode can transfer up to sixty-four words at a time to or from a block transfer (BT) module in a local or remote I/O chassis. Typical BT modules are thermocouple input modules, analog I/O modules, BCD I/O modules, and pulse counters. The block diagram in Figure 7-14 illustrates block transfers from a supervisory PLC5 in scanner mode to a remote I/O rack that has BT modules installed on it. The remote I/O rack has an A-B adapter module (model number 1771-ASB) that communicates internally to the BT modules using the rack data bus.

The A-B PLC5s can also transfer up to sixty-four words at a time between a supervisory process in scanner mode and a processor configured for adapter mode, as shown in Figure 7-15. In this application, both processors simultaneously execute the opposite block transfer instruction.

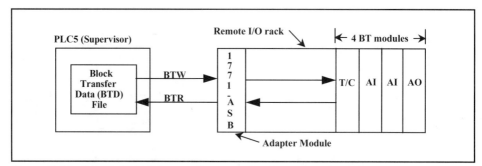

Figure 7-14. A-B PLC5 block transfer operation in scanner mode.

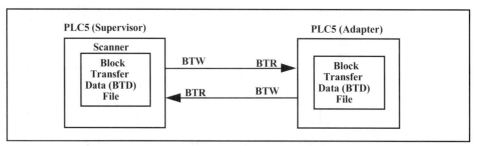

Figure 7-15. A-B PLC5 block transfer operation in adapter mode.

Figure 7-16 shows an example of a bidirectional alternating block transfer. Using rungs like those shown in this example ensures that the block transfer requests are executed in the order in which they were sent to the queue. The processor alternates between the BTR and BTE instructions in the order in which they are scanned by virtue of the enable bit conditions in the rungs. Using the normally closed enabled bits from the two block transfer functions prevents the read-and-write block transfer instructions from queuing simultaneously. In this example, preconditioned normally open instructions are used to the left of the two enable bit contacts. These preconditioned bits allow time-driven or event-driven transfers.

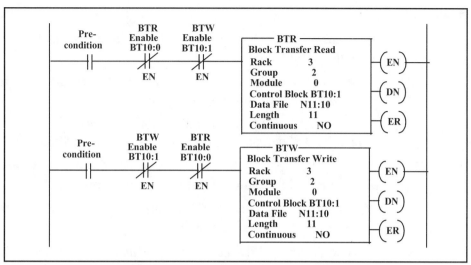

Figure 7-16. A-B PLC5 bidirectional alternating block transfer.

EXERCISES

7.1 List the most common advanced LAD instructions encountered in the Allen-Bradley PLC5 family.

7.2 What is the definition of a file in a PLC memory system?

7.3 What is an Allen-Bradley PLC5 FAL instruction?

7.4 Write an Allen-Bradley PLC5 LAD program to copy the process data located in integer file #N30, words 0, 1, 2, and 3, to integer file #N31, starting at word 0, if input bit I:001/01 is TRUE.

7.5 Write an Allen-Bradley PLC5 LAD program to review the data in integer file #N50, words 0 through 20, and to compare it for an equal condition to the data stored in file #N60, starting at word 1.

7.6 Explain the purpose and function of the control, length, position, mode, destination, and expression in a FAL file instruction.

7.7 What are Allen-Bradley PLC5 LAD shift register instructions used for in industrial applications?

7.8 List the three common sequencer instructions and explain their purpose and use.

7.9 Define the parameter length in a sequencer file.

7.10 What is the position parameter in a sequencer file?

BIBLIOGRAPHY

1.   Allen-Bradley Co. Inc. *Programmable Controller Fundamentals* (Allen-Bradley, 1985).

2.   Bryan, E. A., and L. A. Bryan. *Programmable Controllers: Concepts and Applications* (Industrial Text Co., 1988).

3.   Gilbert, R. A., and J. A. Llewellyn. *Programmable Controllers: Practices and Concepts* (Industrial Training Corp., 1985).

4.   Jones, C. T., and L. A. Bryan. *Programmable Controllers: Concepts and Applications* (International Programmable Controls, 1983).

5.   Rockwell Software, Inc. *PLC-5 Programmable Controller: Instruction Set Reference* (Rockwell Software, 1996).

6.   Reis, R. A., and J. W. Webb. *Programmable Controllers: Principles and Applications*, 3d ed. (Prentice-Hall, 1995).

8

Standard PLC Programming Languages

Introduction

In this chapter, the international standard for PLC programming languages will be discussed. This standard lists five PLC languages: ladder diagram (LAD), function block diagram (FBD), sequential function chart (SFC), statement list (STL), and structured text (ST). Three of these standard languages will be discussed here: sequential function chart, statement list, and structured text. Ladder diagram is the most widely used PLC language was discussed in chapters 6 and 7. Function block diagram language will be covered in the next chapter.

International Standard for PLC Languages

There have been numerous PLC programming standards proposed by different national and international committees to develop a common interface for programmable controllers. Then, in 1979, a working group of international PLC experts was appointed by several national committees to write a first draft of a comprehensive PLC standard. The first committee draft was issued in 1982. After the national committees gave the document an initial review, they decided the standard was too complex to treat as a single document. As a result, the working group was split into five task forces, one for each part of the standard. The subject of each part was as follows: part 1, general information; part 2, equipment and testing requirements; part 3, programming languages; part 4, user guidelines; and part 5, communications. Each task group consisted of several international experts, each backed by a national advisory group. IEC 61131-3, the standard for PLC programming languages, was issued by the International Electrotechnical Commission in March 1993.

The IEC 61131-3 standard has three graphical languages: ladder diagram (LAD), function block diagram (FBD), and sequential function chart (SFC). It also has two text-based languages: statement list (STL) and structured text (ST). The PLC language standard allows different parts of an application to be programmed in different languages and then be combined into a single executable program.

The ladder diagram (LAD) is the most common programmable controller language. It consists of a set of instructions that will perform the most basic types of control functions: relay-type logic, timing and counting, and basic math operations. However, depending on the programmable controller model, the programmer may extend or enhance the instruction set to perform other operations. These additional functions are used for analog control, data manipulation, reporting, complex control logic, and other functions.

The function block diagram (FBD) is a graphical programming language that uses logic blocks similar to those used in Boolean algebra to represent basic logic. It also uses more complex function blocks to perform such operations as timing, counting, math, loading data, transferring data, and comparing data. The programmer is able to build complex control schemes by using functions from the FBD library and then interconnecting them in a graphical diagram area.

Statement list (STL) is a low-level programming language. It is very effective for small simple applications or for optimizing parts of an application. Instructions always relate to the current result, and the operator indicates the operation that must be made between the current value and the operand. The result of the operation is stored again in the current result.

Structured text (ST) is a high-level structured language designed for automation processes. It is used mainly to implement complex procedures that cannot be easily expressed with graphical languages. ST is the default language for describing the actions within the steps and conditions attached to the transitions of the SFC language.

Sequential function chart (SFC) is a graphical language that is used to describe sequential operations. The control process is represented as a set of well-defined steps linked by transitions. A Boolean logic condition is attached to each transition. Actions within the steps and the logic transitions between them can be performed by using instructions from the other standard PLC languages. However, the LAD and STL logic instructions are mainly used to perform the logic transitions between steps.

Sequential Function Chart Language

The SFC language is used to describe operations that are sequential in nature. It uses a simple graphical representation of the different steps of a process and the logical conditions that enable the transitions between active steps. Within an SFC program, instructions from other standard languages are used to describe the actions within the steps and the logical conditions for transitions between steps.

The sample SFC program shown in Figure 8-1 illustrates the symbols used in a typical program. The top of the program contains a step block that is the *initial step*. This is where the programmable controller begins executing the function chart, and it is the step where the controller returns at the end of the program unless it is directed otherwise by the program logic. This block is identified by a double-sided box in Figure 8-1.

The *step* block is the function chart's basic unit. It contains ladder logic for each independent stage of the process or machine operation. It is identified by a single-sided box with the step number inside. The *transition* is the logic condition that the processor checks after completing the active step. When the transition logic is TRUE, the step preceding the transition is disabled, and the step following it becomes active. The transition is normally a single LAD logic rung or STL statement, which is identified by a short horizontal line below its corresponding step. In Figure 8-1, there are six transitions, labeled T1 through T6.

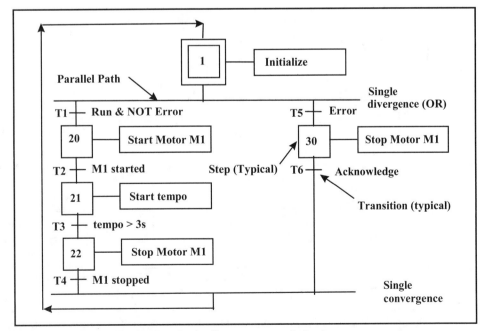

Figure 8-1. Typical SFC with single divergence and single convergence.

The *OR path* is identified by a single horizontal line at the beginning and end of a logic zone, as shown in Figure 8-1. The processor selects one of several parallel paths depending on which transition goes TRUE first. The *AND path* is identified by a horizontal double line at the beginning and end of a zone, as shown in Figure 8-2.

In sequential function chart programs, steps and transitions are arranged in series and parallel paths, and they are numbered with the file numbers that contain their logic. The programmable controller scans the logic of a step repeatedly until its transition logic goes TRUE. Then the program scan moves to the next step or steps, and the previously active step is turned OFF.

There are three basic rules for standard sequential function chart programs. The first is that the *initial* step is always activated at start-up. When the programmer is restarting from the beginning of the chart as well as on subsequent passes through the flowchart program, the programmer does not have the option of restarting from the beginning of the last active step(s). Nor does the programmer have the option of following or changing the programmable controller's mode from run to program and back to run again.

The second rule for standard sequential function chart programs is that a transition is tested after its associated step and operations pass from one step to the next through a transition when the transition goes TRUE. The

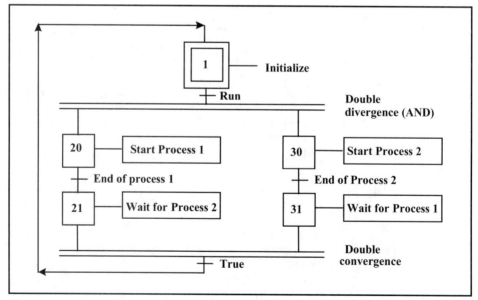

Figure 8-2. Typical SFC program with double divergence and convergence.

third rule is that after a TRUE transition the processor scans the step once more to reset all timer instructions and then executes the next step. This extra processor scan is called *postscan*. It is important to note that the processor never postscans a transition file, so timers probably should not be used in transition files.

The processor scans a sequential function chart program from left to right and top to bottom. When the SFC program scan encounters active parallel steps, it executes the ladder logic in the left-most step first. It then moves across the screen from left to right to the ladder logic in the next parallel step.

SFC Application

To illustrate an SFC application, consider a simplified example of a semiautomatic punch (see Figure 8-3). Assume that the punch starts in the raised or top position. When the operator depresses the start pushbutton, the punch is lowered, and it pierces the metal part at the lowest or bottom position. The cycle is completed when the control system raises the punch back to the top position and the operator removes the punched metal part and inserts a new piece of metal for the next operation.

The semiautomatic punch control system shown in Figure 8-3 has four process states or steps. The first process step is the punch at rest in the raised position. The second step is the punch at rest waiting for a fault to be acknowledged. The third step is the punch descending, and the final step is the punch ascending.

The SFC program shown in Figure 8-4 controls the operation of the punch. Its first process step is labeled "1." When the punch and associated control system are activated, the programmable controller places the system in this wait state (step 1) and waits for the first transition logic (T1) to be satisfied. If the start pushbutton (PB) is depressed *and* the punch is in the

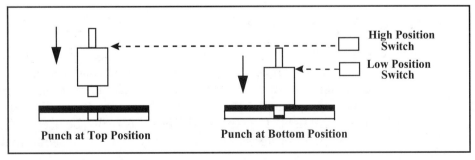

Figure 8-3. Semiautomatic metal punch example.

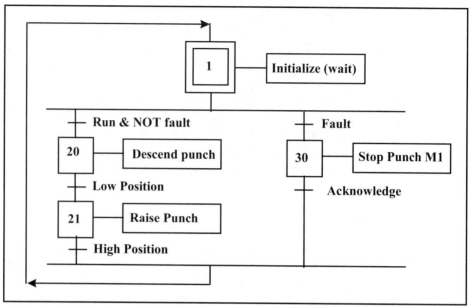

Figure 8-4. SFC program for semiautomatic metal punch control.

top position *and* there is no fault present, the first transition is TRUE and step 20 is activated. As a result, the punch moves down. At the same time, the programmable controller turns OFF the first state.

When the punch activates the bottom limit switch, the second transition (T2) becomes TRUE and the punch raises (step 21). Finally, when the punch reaches the top, the T2 transition bit is set, and the control program returns to the wait state (step 1).

Step 30 (punch stopped) occurs when the punch is activated but there is a fault present. In this case, the third transition (T3) is TRUE and step 30 is activated. The punch is stopped, waiting for the fault to be acknowledged. If the fault is acknowledged, the control program returns to the initialized state (step 1) and waits for a new start command.

Structured Text Language

In this section, structured text (ST), a high-level language designed for complex automation processes, will be covered. This language is used mainly to implement complex procedures that cannot be easily expressed with graphical languages, such as LAD and function block diagram or the simpler statement list (STL) language. The ST can also be used to implement the process steps and transition conditions in sequential function chart programs.

An ST program is a list of programming statements. Each statement ends with a semicolon (;) separator. Names used in the source code, such as variable identifiers, constants, and language keywords, are separated by inactive separators (space character, end-of-line character, or stops). Or they are separated by active separators that have a well-defined significance (for example, the ">" separator indicates a greater-than comparison). The programmer can freely insert programming comments into a program list for clarity, but a comment must begin with the two characters (* and end with the two characters *). Each statement in the program terminates with a semicolon (;) separator character.

The following are basic statement types for ST programs:

1.   *Assignment* statement (for example, *variable:=expression;*)

2.   *Subprogram or function* call

3.   *"C" function block* call

4.   *Selection* statements (IF, THEN, ELSE, CASE, etc.)

5.   *Iteration* statements (FOR, WHILE, REPEAT, etc.)

6.   *Control* statements (RETURN, EXIT, etc.)

7.   *Special* statements for links with other languages, such as SFC

The programmer may freely insert inactive separators between active separators, constant expressions, and identifiers to improve readability. The ST inactive separators are *space* (blank character), *tabs*, and *end-of-line* characters. Unlike line-formatted languages such as statement list (STL), end–of-lines may be entered anywhere in the program. This greatly increases the readability of the program. Follow these three programming tips to increase ST program readability when using inactive separators:

1.   Do not write more than one statement on one line.

2.   Use tabs to indent complex statements.

3.   Insert comments to increase the readability of lines.

The left side of Figure 8-5 shows an ST program with low readability. The right side shows the same program with high readability. This simple example shows how to improve the readability of an ST program.

Low readability ST Program	Same ST program with high readability
imax:=max_ite;cond:=X12 if not(cond(*alarm*)) then return;end_if; for i (*index*):=1 to max_ite end_if;end_for; (*no effect if alarm*)	(*imax: number of iterations*) (*i: FOR statement of index*) (*cond: process validity*) imax:=max_ite; cond:=X12 if not(cond(*alarm*)) then return; end_if (*process loop*) for i (*index*):=1 to max_ite do if i<>2 then SPcall(); end_if;end_for; end_for;

Figure 8-5. Example of ST program readability.

ST Expressions and Parentheses

ST expressions combine operators and variables or constant operands. For each single expression (combining operands with one ST operator), the type of the operands must be the same. The following are examples of valid and invalid expressions:

1.   *(boo_var1 AND boo_var2).* This expression is valid because there are two Boolean variables (*boo_var1* and *boo_var2*) being operated on by a Boolean AND operand, so all items in the expression are the same type.

2.   *NOT(boo_var1).* This is a valid expression because there is a Boolean variable (*boo_var1*) being operated on by a Boolean NOT operand.

3.   *(1s23 + 1.78).* This is not a valid expression because it uses a mixture of types.

Parentheses are used to isolate parts of the expression and to explicitly order the priority of the operations. If no parentheses are used in a complex expression, the operational sequence is implicitly given by the default priority between ST operators. For example, *2+3*6* equals *2+18=20* because the multiplication operator has a higher priority than the addition operator. However, if parentheses are used on the same equation, the result of the operation on *(2+3)*6* is *5*6=30* because the parentheses give priority to the addition operation.

Statement List Programming

Statement list (STL) is a textual programming language that can be used to create the code for a PLC control program. Its syntax for statements is similar to microprocessor assembly language and consists of instructions followed by addresses on which the instructions act. The STL language contains a comprehensive range of instructions for creating a complete user program. For example, in the Siemens S7 programming software package, over 130 different basic STL instructions and a wide range of addresses are available, depending on which model PLC is used.

In this section, the statement list language that is used to program Siemens S7-300 and S7-400 programmable controllers will be discussed. Most other PLC manufacturers use a similar version of STL. This section will provide the basics of Siemens STL programming along with some typical applications. However, the programmer should consult the Siemens reference manual for Simatic S7 statement list (STL) programming for detailed information on how to use the software.

STL Statement Structure

STL instruction statements have two basic structures. The first is a statement made up of an instruction alone (for example, NOT), and the second is a structure in which the statement is made up of an instruction and an address. This second case is the most common structure.

The address of an instruction statement indicates a constant or a memory location in which the instruction finds a value on which to perform an operation. The address can have a symbolic name or an absolute memory designation. The address can point to a number of items, such as a constant, a bit in a status, a symbolic name, a memory data block, and so on.

Table 8-1 shows a constant value (+77) and a character string (END) being used as the address of an instruction. The first instruction loads the constant value or integer 77 into accumulator 1 in the central processing unit (CPU) of the PLC. The Siemens S7 series programmable controllers have two 32 bit accumulators. These accumulators are general-purpose registers that are used to process bytes, words, and double words. They load constants or values from PLC memory as addresses and perform logic operations on them. The control program can also transfer the results of an operation from the first accumulator (accumulator 1) to a PLC memory location. The structure of accumulator 1 or 2 is given in Figure 8-6. Bits 0 through 7 are the lower byte of the lower word, and bits 8 through 15 are the upper byte (High Byte) of the lower word (Low Word). The High Word is bits 16 through 31. It has the same Low and High Bytes as shown in Figure 8-6. The second instruction loads the ASCII character "END" into accumulator 1. At the same time, the CPU moves the integer 77 into accumulator 2.

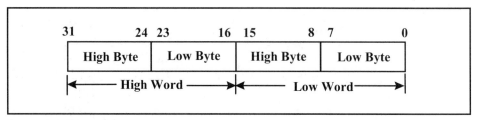

Figure 8-6. Word and byte structure for accumulators.

The address of a STL instruction can also refer to one or more bits in the status word of the programmable controller. The two instructions listed in Table 8-1 are examples of an instruction operating on status word bits. The "status" word structure is shown in Figure 8-7. Bit cell 0 of the status word contains the "First Check" (FC) bit. The bar over the FC in bit 0 indicates that the bit is negated.

Table 8-1. Examples of Constant Values as an Address

STL Instruction	Description
L +77	Load the integer 77 into accumulator 1.
L "END"	Load the ASCII character "END" into accumulator 1.

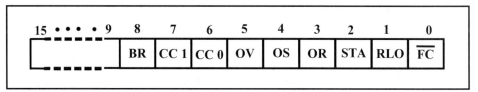

Figure 8-7. Status word structure.

Bit cell 1 contains the "Result of the Logic Operation" (RLO) bit. This bit cell stores the result of a bit logic instruction or math comparison. Bit cell 2 contains the "Status" (STA) bit. It stores the value of a bit that is referenced in a program. Bit cell 3 stores the "Or Operation" (OR) status bit. The "Overflow" (OV) bit indicates a fault encountered during the execution of a math instruction or a floating-point comparison instruction. The "Stored Overflow" (OS) bit is set together with the overflow (OV) bit when a fault occurs. The "CC 1" and "CC 0" bits (condition codes) provide information on the following: (1) the results of a math operation, (2) the results of a comparison, (3) the results of a digital operation, or (4) the use of a shift or rotate command to shift bits out. The "Binary Result" (BR) bit forms a link between the processing of bits and words.

Bit Logic Instructions

The basic type of STL instructions are the bit logic instructions. These instructions perform logic operations on single bits in PLC memory. The basic bit logic instructions are **AND** (A) and its negated form **AND NOT** (AN); **OR** (O) and its negated form **OR NOT** (ON); and **EXCLUSIVE OR** (XO), and its negated form **EXCLUSIVE OR NOT** (XN). These instructions check the signal state of a bit address to establish whether the bit is activated (logic 1) or not activated (logic 0). They can also be used to check the status of the count or time value in a timer or a counter instruction. If the time or count value is greater then zero, the Result of the Logic Operation (RLO) bit is set to one (1). If the time or count value is zero, the RLO bit is set to zero.

AND Logic Operation

The result of an *AND logic operation* is logic 1 only if all inputs are logic 1. Figure 8-8 is an example of **AND** logic operation with two inputs. The STL program is listed on the left side, and the Ladder Diagram (LAD) program is shown on the right side for comparison. The statement list (STL) program uses an AND (A) on the two binary input bits at I124.0 and I124.1. If both bits are 1, then the output bit Q124.0 is set to 1. Figure 8-8 shows the ladder diagram program for comparison. With two normally open instructions in series, when the logic state of both the input bits is 1 the output coil is 1. This is the same AND function as the statement list program on the left side of Figure 8-8.

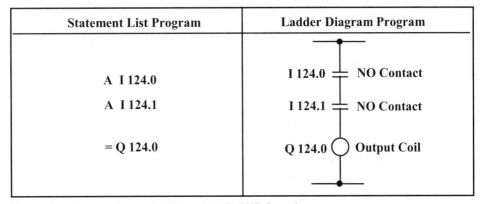

Statement List Program	Ladder Diagram Program
A I 124.0	I 124.0 — NO Contact
A I 124.1	I 124.1 — NO Contact
= Q 124.0	Q 124.0 ◯ Output Coil

Figure 8-8. Comparison of STL and LAD AND function programs.

OR Logic Operation

The result of an *OR logic operation* is logic 1 if any or all inputs are logic 1. Figure 8-9 is an example of an **OR** logic operation on two inputs. The STL program is listed on the left side, and the ladder diagram (LAD) program is shown on the right side to compare the same program in the two different programming languages. In this example, the statement list program uses two **OR** (O) instructions to perform a logical **OR** operation on inputs I124.0 and I124.1. To perform an **OR** logic operation, the LAD program uses two normally open (NO) contact instructions in parallel and connected to an output coil. In both programs, when the logic state of one or both of the inputs is 1, the output bit, Q124.0, is to set to logic 1. If both inputs are 0, the output is set to 0.

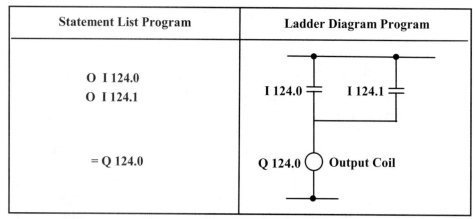

Statement List Program	Ladder Diagram Program
O I 124.0 O I 124.1 = Q 124.0	

Figure 8-9. Comparison of STL and LAD OR function programs.

EXCLUSIVE OR Logic Operation

The result of an *EXCLUSIVE OR logic operation* is logic 1 only if one of the inputs is 1, but the result is 0 if both inputs are logic 0 or 1. The use of an **EXCLUSIVE OR** instruction in a STL program is shown on the left side of Figure 8-10. The corresponding circuit in a ladder diagram (LAD) program is shown on the right side of the figure. In these sample programs, the output Q124.0 is logic 1 only if inputs I124.0 and I124.1 have different logic values. The output is 0 if the two inputs have the same logic value.

The equivalent two-input **EXCLUSIVE OR** LAD program is shown on the right-hand side of Figure 8-10 for comparison. In this LAD program, the output coil bit Q124.0 is energized only if the input bits have different values.

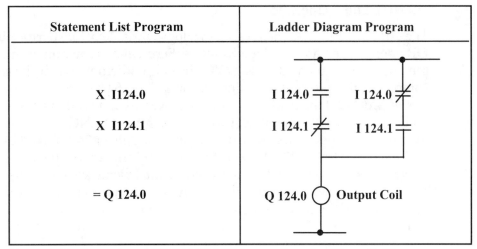

Statement List Program	Ladder Diagram Program
X I124.0	
X I124.1	
= Q 124.0	

Figure 8-10. Examples of EXCLUSIVE OR instructions in STL and LAD programs.

AND NOT Logic Operation

The *AND NOT logic operation* negates the logic input before performing an **AND** operation with another logic input. Figure 8-11 shows the use of the statement list **AND NOT** (AN) instruction. The STL program is listed on the left side of the figure, and the equivalent ladder diagram (LAD) program is shown on the right side for comparison. In this example, the statement list program uses **AND NOT** (AN) instructions on input bits I124.0 and I124.1. The output is set to logic 1 only if both input bits are logic 0. This is the same logic operation as performing an **AND** operation on two normally closed relay contacts in a relay ladder logic circuit, as shown in Figure 8-11.

Statement List Program	Ladder Diagram Program
AN I 124.0	I 124.0 — NC Contact
AN I 124.1	I 124.1 — NC Contact
= Q 124.0	Q 124.0 ◯ Output Coil

Figure 8-11. Example of the use of STL AN instructions.

OR NOT Logic Operation

The *OR NOT logic operation* negates its logic variable before performing an **OR** operation on another logic variable. Figure 8-12 shows an example of the use of a statement list **OR NOT** (ON) logic instruction. For the sake of comparing the two programming languages, the STL program is listed on the left side, and the equivalent ladder diagram (LAD) program is shown on the right. The statement list program uses two **OR NOT** (ON) instructions to perform a logic operation on inputs I 124.0 and I 124.1, to activate the output Q124.0. To perform the logic operation, the LAD program uses two normally closed (NC) contact instructions in parallel and connected to an output coil. In both programs, when the logic state of one or both of the two input bits is 0, the value of output bit Q124.0 is set to 1.

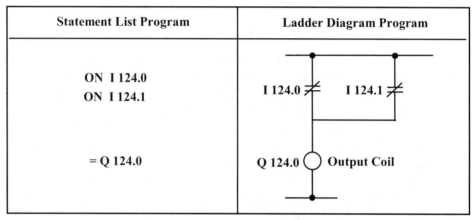

Figure 8-12. Use of the OR NOT (ON) statement list instruction.

Boolean logic operations can be performed on the portions of a logic string that are enclosed in parentheses. These logic operations are called *nesting expressions*. Parentheses around a portion of a logic string mean the program will perform the operations inside the parentheses before performing the logic operation indicated by the instruction that precedes the nesting expression.

AND and **OR** statements can also be combined in a Boolean logic string without using parentheses. By convention, the **AND** statements are evaluated first, and the results are then combined according to the **OR** truth table.

Timer Instructions

Software timers provide the same functions as hardware timers in process control applications. A typical application for a software timer instruction is to delay an operation for a fixed time interval. For example, the starting of a pump might be delayed for several seconds until a valve on the discharge line of the pump is completely opened so high pressure cannot develop at the discharge of the pump. Another example would be to have a timer determine the amount of time a vessel takes to fill with fluid and then display the time to the plant operator on a graphic display.

Timer Word Structure

The PLC memory area reserves one 16-bit word for each timer address in a timer instruction. The Siemens Simatic S7 statement list language instruction set supports up to 256 timers. However, the exact number of timers supported depends of the model number of the CPU used in the application.

The STL timer instruction in a Siemens S7 PLC uses bits 0 through 11 of accumulator 1's Lower Word to hold the time value. The programmer can load a time value into the Lower Word of accumulator 1 in binary, hexadecimal, or binary-coded decimal (BCD). The BCD format for the time value is shown in Figure 8-13. The left-most two bits (bits 14 and 15) in the timer word are irrelevant and are ignored when the timer is started. Figure 8-13 shows the timer word loaded with a timer value of 349 and with a time base of one second. The 12 bits in cell locations 0 through 11 are divided into three groups of four bits each to produce a three-digit BCD number. Bits 12 and 13 hold the time base in binary code.

There are four time bases available for Simatic S7 timers: 10 ms, 100 ms, 1 second, and 10 seconds. Table 8-2 lists the time bases and their corresponding binary code. The time range is from 0 to 9,990 seconds.

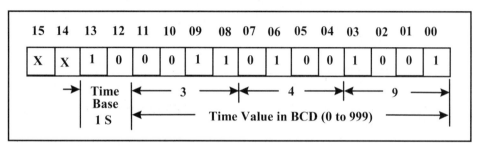

Figure 8-13. Timer memory word structure in accumulator 1.

Table 8-2. Time Base and Its Binary Code

Time Base	Binary Code for Time Base
10 ms	00
100 ms	01
1 second	10
10 seconds	11

Because time values are stored in memory with only one time interval, time values that are not exact multiples of a time interval are truncated. Time values that have too much resolution for the desired range are rounded down to achieve the desired range but not the desired resolution. Table 8-3 lists the possible resolutions and their corresponding time ranges.

Table 8-3. Time Base Resolution and Ranges

Resolution	Range
0.01 second	10MS to 9S_990MS
0.1 second	100MS to 1M_39S_900MS
1 second	1S to 16M_39S
10 seconds	10S to 2HR_46M_30S

The time base and the time value are loaded in a STL program using a **LOAD** (L) statement. Two syntax formats can be used to load a time value. The first format is *L W#16#wxyz*, where *w* is the time base and *xyz* is the time value in BCD. The other format is *S5T#aH_bbM_ccS_dddMS*, where *aa* is time value in hours, *bb* is a time value in minutes, *cc* is a time value in seconds, and *ddd* is a time value in milliseconds. In this second method, the software automatically selects the time base, and the value is rounded to the next lower number with that time base. For example, to program a timer for ten seconds, type in the list statement instruction "L S5T#10S0MS." The software would select a time base of 0.1 second, according to Table 8-3. This time base resolution is selected because the time interval of 10 seconds is between the range of 100 milliseconds and 1 minute, 39 seconds and 900 milliseconds, which gives the lower or best resolution possible.

Programming an STL Timer

A minimum of three instructions must be used to program for the **Timer** function in a Siemens S7 PLC using STL programming language. First an instruction to check the status of a timer start bit must be listed in the program, for example, "A I124.1" as shown in the sample program in

Figure 8-14. Next, an instruction to load a time value must be included in the program, for example, "L S5T#00H02M10S00S" as shown in Figure 8-14. Finally, an instruction for the type of timer to be used in the application must be listed in the program. There are five different types of timers available in the statement list language set: pulse timer (SP), extended pulse timer (SE), on-delay timer (SD), retentive on-delay timer (SS), and off-delay timer (SF).

In a statement list program, a change in the logic value of the timer start bit prior to a start instruction statement starts a timer. A change in the start bit from 1 to 0 starts an off-delay timer (SF), and a logic change of 0 to 1 in the start bit starts any of the other timers. The programmed time and the start timer statements must directly follow the logic operation that provides the condition for starting the timer. This is shown in the simple example in Figure 8-14.

Because a timer runs down to zero from a set time, a STL program must provide the timer with a starting time. For example, a starting time of one minute and ten seconds is used in the STL timer program shown in Figure 8-14.

STL Program	Description
A I124.1	Check the logic state of "Start" bit I124.1
L S5T#1M10S	Load starting time of 1 minute & 10 seconds
SP T1	Start timer T1 as a pulse timer

Figure 8-14. Example of an STL timer program.

A STL timer can be reset by using a reset (R) instruction in the program. The CPU resets a STL timer when the result of a logic operation is one just before the R instruction in the program. Resetting a timer stops the current timing function and resets the time value of the timer to zero.

An **Enable** instruction (FR) can be used to restart the timing interval of a STL timer. A change in the state of the logic bit just before the FR instruction from logic 0 to logic 1 will restart a running timer. A timer **Enable** is not required to start a timer, nor is it required for normal operation. The next section on pulse timers will provide an example of how to use an **Enable** instruction.

STL Timer Application

A statement list timer application program using a pulse timer is shown in Figure 8-15. The program uses enable, start, and reset bit inputs to control the operation of a pulse timer.

When the start bit shown in the timing diagram of Figure 8-15 changes from a logic state of 0 to a logic state of 1 at time t_0, the pulse timer will start. The programmed time then runs down to zero at time t_1. At time t_2, the start bit goes from logic 0 to logic 1, starting the timer. At time t_3, the reset bit changes from logic 0 to 1 and the timer is reset to zero. The timer start bit is returned to zero at time t_4; then at time t_5 the start bit returns to

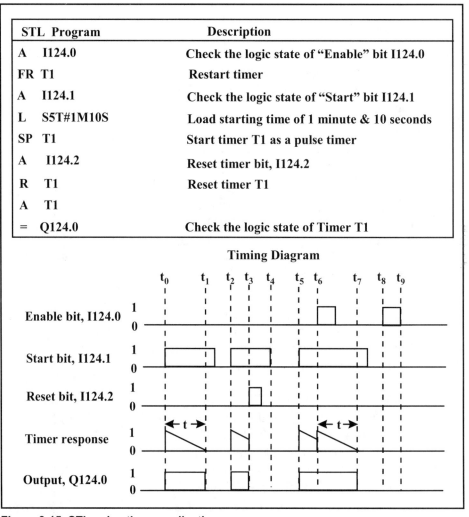

STL Program	Description
A I124.0	Check the logic state of "Enable" bit I124.0
FR T1	Restart timer
A I124.1	Check the logic state of "Start" bit I124.1
L S5T#1M10S	Load starting time of 1 minute & 10 seconds
SP T1	Start timer T1 as a pulse timer
A I124.2	Reset timer bit, I124.2
R T1	Reset timer T1
A T1	
= Q124.0	Check the logic state of Timer T1

Figure 8-15. STL pulse timer application.

one (1), starting the timer. At time t_6, the enable bit goes from logic 0 to logic 1 with the timer running. This restarts the timer. The timer runs until time t_7, and the time is extended by the amount of the programmed time interval. At time t_8, the enable bit again goes from logic 0 to logic 1 until time t_9, but there is no effect on the timer because the start bit is OFF and the timer is not running.

If an application requires that the timer output be delayed by a fixed amount of time, an on-delay (SD) timer instruction can be used. A typical application of an SD timer instruction is shown in Example 8-1.

EXAMPLE 8-1

Problem: Write an STL program for a Siemens Simatic S7-300 PLC that delays the starting of a process pump for ten seconds to allow a valve in the discharge line of the pump to fully open. Assume that the pump starter relay is wired to PLC output point Q124.2 and that an operator uses a normally open switch connected to input point I124.0 to start the pump.

Solution: The following is the STL program to operate the process pump, with an explanation of each instruction:

STL Instruction	Explanation
A I124.0	Check logic of Pump Run input.
L S5T#10S	Load 10-second time delay if Pump Run bit is 1.
SD T2	Start timer T2 as an "On-delay" timer.
A T2	Check timer output status.
=Q124.2	Start process pump after 10-second delay.

Counter Instructions

Counters are instructions that provide the same functions as hardware counters in process control applications. In some applications, they are used to activate or deactivate a control device after a set count has been reached. For example, a STL control program might be used to count the number of parts produced on an assembly line and then stop the production line after a given number of parts have been manufactured.

There are two counter instructions in the STL programming language: **Count Up** (CU) and **Count Down** (CD). An application that uses both types of counter instructions will be covered, but first the structure of the STL counter word will be discussed.

Counter Word Structure

The memory area of a PLC reserves one 16-bit word for each counter in a counter instruction. The Siemens Simatic S7 statement list language instruction set supports up to 256 counters, but the exact number of counters supported depends on the model number of the CPU used in the application.

The STL counter word in a Siemens S7 PLC uses bits 0 through 9 to hold the count value in binary format. A **LOAD** (L) instruction can be used to read the binary count value out of a counter word and load the count value into the Low Word of the accumulator. For example, the STL instruction "L C2" will directly load accumulator 1 low with the count value of counter number 2 in binary format. A **TRANSFER** (T) instruction can be used to transfer the count value to another location memory so it can be used in a control application. A load counter (LC) value instruction can be used to read the binary count value out of a counter word and to load the count value into the Low Word of accumulator 1 in binary-coded decimal (BCD) format. For example, the STL instruction "LC C3" will directly load accumulator 1 low with the count value in BCD format from counter number 3, as shown in Figure 8-16.

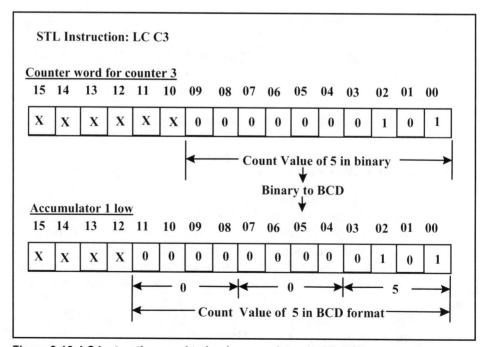

Figure 8-16. LC instruction used to load accumulator 1 with BCD count value.

Counter Application

The STL programming application shown in Figure 8-17 provides an example of counting up, counting down, setting and resetting a counter, checking the logic state of a counter, and loading a fixed count value into a counter. In the first line of the STL program (A I124.6), the logic state of the Count Up bit, I124.6 is checked. If this bit is set to logic 1, the STL instruction CU C1 in line 2 will increment the counter by one count.

STL Program	Description
A I124.6	Check the logic state of "Count Up" bit I124.6
CU C1	Increment count value by 1
A I124.7	Check the logic state of "Count Down" bit I124.7
CD C1	Decrement count value by 1
A I124.2	Check the logic state of "Set Counter" bit I124.2
L C#3	Load count value of 3
S C1	Set count value to 3 in counter number 1.
A I124.3	Check logic state of "Reset Counter" bit I124.3
R C1	Reset count value to 0
A C1	Check the logic state of Counter C1
= Q124.1	Output the logic state of Counter C1

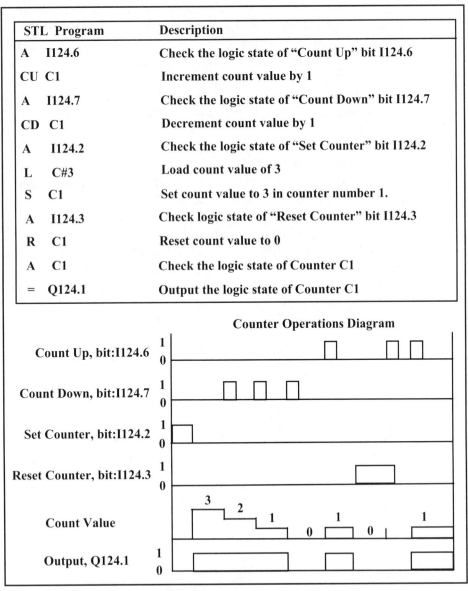

Figure 8-17. Up/down counter programming application.

However, if the reset counter bit I124.3 is one (1) at the same time, the count value will not be incremented by one. The third line of the program (A I124.7) examines the logic state of the Down Count bit, I124.7. If this bit is set to one (1), the instruction DU C1 in line 4 of the program will decrement the counter by one count. The fifth line of the program (A I124.2) examines the logic state of the Set Counter bit I124.2. If this bit is 1, a count value of 3 is loaded into counter 1 by the instructions LC#3 and S C1 in lines 6 and 7, respectively, in the program.

In the counter operation diagram shown at the bottom of Figure 8-17, the set counter bit is logic 1, and a count valve of 3 is loaded into counter 1. Later, the Count Down bit is changed from zero to one and then back to zero, three consecutive times. Each time, the Count Down bit goes from zero to one the counter is decremented by one. This returns the count value to zero. Next, the Count Up bit is changed from zero to one and then back to zero, and the counter is incremented by one on the logic change of zero to one.

The last two STL instructions (A C1 and = Q124.0) are used to check the status of the count value in the counter. If the count value is zero, the output logic bit Q124.1 is set to one. In all other cases, the bit is set to zero.

In the counter application shown in Figure 8-17, a second Count Up pulse occurred at the same time as the count reset bit is one (1). In this case, the counter value is momentarily incremented by one, but it is immediately reset because of the reset instruction that follows directly in the program. This momentary increase is indicated by a pulse line in the counter operations diagram shown in Figure 8-17.

A typical control application using a STL counter instruction is given in Example 8-2.

Integer Math Instructions

In this section, STL instructions used to perform mathematical operations on positive and negative whole numbers (1, 2, 3, etc.) or zero (i.e., integers) will be discussed. The basic STL math instructions used to add, subtract, multiply, and divide single word (16-bit) integers and double word (32-bit) integers are listed in Table 8-4.

EXAMPLE 8-2

Problem: Write a STL program for a Siemens Simatic S7-300 PLC to turn off a conveyor belt on a production line after 150 parts have been produced. Assume the following: (1) output bit Q124.3 = 0 turns off the conveyor belt; (2) input bit I124.7 changes from 0 to 1 and then back to 0 each time a new part is produced; (3) a normally open (NO) pushbutton connected to input I124.2 is used to set the production count to 150, and (4) A NO pushbutton connected to input I124.3 is used to reset the counter to zero and stop the conveyor belt.

Solution: The following shows the STL program to control the operation of the conveyor belt, with an explanation for each instruction:

<u>STL</u> Instruction	<u>Explanation</u>
A I124.7	Check "Part Produced" input for logic 1.
CD C2	Decrement count by 1 if a part is produced.
A I124.2	Test "Set part count" bit I124.2 for logic 1.
L C#150	Load count of 150.
S C2	Set count value to 150 for counter number 2.
A I124.3	Check logic state of "Reset Counter" input.
R C2	Reset count to 0, if I124.3 = 1.
A C2	Check logic state of counter C2.
= Q124.3	Run conveyor if count value is not 0.

Table 8-4. Integer Math Instructions

STL Instruction	Description
+I	Add Single Word Integer
+D	Add Double Word Integer
-I	Subtract Single Word Integer
-D	Subtract Double Word Integer
*I	Multiply Single Word Integer
*D	Multiply Double Word Integer
/I	Divide Single Word Integer
/D	Divide Double Word Integer

Integer Add Instructions

The Add Single Integer (+I) instruction adds the contents of the Low Words of accumulators 1 and 2 and stores the result in the Low Word of accumulator 1. This operation overwrites the old contents of the Low Word of accumulator 1. The old contents of accumulator 2 and the High Word of accumulator 1 remain unchanged, as shown in Figure 8-18.

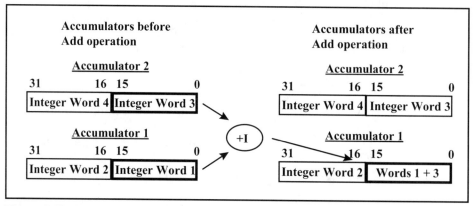

Figure 8-18. Accumulator operation for an integer add instruction.

Figure 8-19 shows an example of a STL program using an Add Integer instruction. The first instruction, L MW10, loads the integer value in memory word MW10 into accumulator 1. The second instruction, L MW12, loads the integer value in data word MW12 into accumulator 1. The CPU shifts the old contents of accumulator 1 into accumulator 2. The third instruction, +I, tells the CPU to add the Low Words in accumulator 1 and 2 and store the result in the Low Word of accumulator 1. The last instruction, T MW14, instructs the CPU to transfer the result of the addition that is in the Low Word of accumulator 1 into memory word DBW14.

STL Instructions	Description
L MW10	Load the integer value in MW10 into accumualtor 1
L MW12	Load the integer value in MW12 into accumualtor 1
+I	Add single word MW12 to single word MW10
T MW14	Transfer integer result to data word MW14

Figure 8-19. Typical STL program using Integer Add instruction.

Constant single integers can be added to another integer by first loading the integer constant into the Low Word of accumulator 1. This moves the old value in the Low Word of accumulator 1 into the Low Word of accumulator 2. Then an Add instruction can be used to add the Low Words of accumulators 1 and 2. For example, to add the whole number 5 to the value in accumulator 1, the instruction L 5 is used. This instruction loads whole number 5 into the Low Word of accumulator 1. Then, an Add (+I) instruction can be performed. The result of this add is stored in accumulator 1, overwriting the old contents of the Low Word in accumulator 1. The content of accumulator 2 does not change after the Add operation is performed.

Constant double word integers can be added to the contents of accumulator 1 by using the LOAD instruction followed by an add double integer (+D) instruction. For example, to add the number 155 to the contents of accumulator 1, we first use the instruction +L#155. The number 155 is loaded into accumulator 1, and the old contents of accumulator 1 are moved to accumulator 2. Then, an Add Double Integer (+D) instruction would add the contents of accumulators 1 and 2 and store the result in accumulator 1. This operation overwrites the old contents of accumulator 1. The old contents of accumulator 2 remain unchanged.

Integer Subtract Instructions

The Subtract Single Integer (-I) instruction subtracts the contents of the Low Word of accumulator 1 from the Low Word of accumulator 2 and stores the result in the Low Word of accumulator 1. This operation overwrites the old contents of the Low Word of accumulator 1. The old contents of accumulator 2 and the High Word of accumulator 1 remain unchanged.

The Subtract Double Integer (-D) instruction subtracts the contents of accumulator 1 from the contents of accumulator 2 and stores the result in accumulator 1. This operation overwrites the old contents of accumulator 1, but the contents of accumulator 2 remain the same.

Integer Multiply Instructions

The Multiply Single Integer (*I) instruction multiplies the contents of the Low Word of accumulator 1 by the contents of the Low Word of accumulator 2 and stores the result in the Low Word of accumulator 1. This operation overwrites the old contents of the Low Word of accumulator 1. The old contents of accumulator 2 and the High Word of accumulator 1 remain unchanged.

The Multiply Double Integer (*D) instruction multiplies the contents of accumulator 1 by the contents of accumulator 2 and stores the result in accumulator 1. This operation overwrites the old contents of accumulator 1, but the contents of accumulator 2 remain the same.

The following example is a typical STL program using integer math instructions that might be encountered in a PLC control application.

EXAMPLE 8-3

Problem: Write a Siemens Simatic S7 STL program to add the integer data in word MW22 to the integer data in word MW40 and then multiply the result by five. Store the result in word MW50.

Solution: The STL program to perform the math operation is as follows:

STL Instruction	Explanation
L MW22	Load the integer value in MW22 into accumulator 1.
L MW40	Load the integer value in MW40 into accumulator 1.
+I	Add value in MW22 to value in MW40.
L 5	Load integer 5 into accumulator 1.
*I	Multiply result of addition (MW22 + MW40) by 5.
T MW50	Store result in MW50.

Integer Divide Instructions

The Divide Single Integer (/I) instruction divides the contents of the Low Word of accumulator 2 by the contents of the Low Word of accumulator 1 and stores the result in the Low Word of accumulator 1. Any whole remainder is stored in the High Word of accumulator 1. This operation overwrites the old contents of accumulator 1, but the contents of accumulator 2 remain unchanged.

The Divide Double Integer (/D) instruction divides the contents of accumulator 2 by the contents of accumulator 1 and stores the result in accumulator 1. This operation overwrites the old contents of accumulator 1, but the contents of accumulator 2 remain the same.

Floating-Point Math Instructions

In this section, the most common STL instructions used to perform mathematical operations on real numbers are discussed. The basic floating-point math instructions—add, subtract, multiply, and divide—are listed in Table 8-5. Consult the Simatic reference programming manual for a complete list of floating-point math instructions available in the S7 PLCs.

Table 8-5. STL Real Math Instructions

STL Instruction	Description
+R	Add 32-bit floating-point numbers
-R	Subtract 32-bit floating-point numbers
*R	Multiply 32-bit floating-point numbers
/R	Divide 32-bit floating-point numbers

Add Real Instruction

The Add Read (+R) instruction adds the 32-bit floating-point numbers in accumulators 1 and 2 and stores the result in accumulator 1. This operation overwrites the old contents of accumulator 1. The old contents of accumulator 2 remain unchanged, as shown in Figure 8-20.

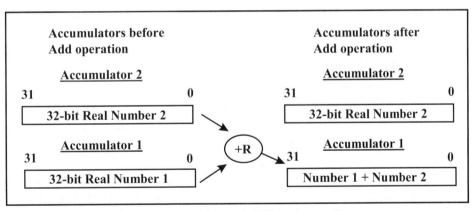

Figure 8-20. Accumulator operation for an Add Real instruction.

Figure 8-21 shows an example of a STL program using an Add Real instruction. The first instruction, L MD100, loads the real number in memory double word MD100 into accumulator 1. The second instruction, L MD104, loads the real number in data double word MW104 into accumulator 1. The old contents of accumulator 1 are shifted into accumulator 2. The third instruction, +R, tells the CPU to add the 32-bit floating-point numbers in accumulators 1 and 2 and to store the result in

STL Instruction	Description
L MD100	Load the real number in MD100 into accumualtor 1
L MD104	Load the real number in MD104 into accumualtor 1
+R	Add real number in MD100 to real number in MD104
T MD108	Transfer the result to double data word MD108

Figure 8-21. Typical STL program using an Add Real instruction.

accumulator 1. The last instruction, T MD108, instructs the CPU to transfer the result in accumulator 1 to double word MD 108.

Subtract Real Instruction

The Subtract Real (-R) instruction subtracts the contents of accumulator 1 from the contents of accumulator 2 and stores the 32-bit result in accumulator 1. This operation overwrites the old contents of accumulator 1. The old contents of accumulator 2 remain unchanged.

Multiply Real Instruction

The Multiply Real (*R) instruction multiplies the contents of accumulator 1 by the contents of accumulator 2 and stores the result in accumulator 1. This operation overwrites the old contents of accumulator 1. The old contents of accumulator 2 remain unchanged.

Divide Real Instructions

The Divide Real (/R) instruction divides the contents of accumulator 2 by the contents of accumulator 1 and stores the result in accumulator 1. This operation overwrites the old contents of accumulator 1, but the contents of accumulator 2 remain unchanged.

The following example is a typical STL program using integer math instructions that might be encountered in a PLC control application.

Comparison Instructions

In this section, the three types of Siemens S7 STL language comparison instructions: Compare Single Integer (I), Compare Double Integer (D), and Compare Real (R) will be covered. Table 8-6 lists the type of comparisons made along with the relational operator used in the STL instruction.

EXAMPLE 8-4

Problem: Write a Siemens Simatic S7 STL program to add a 32-bit floating-point number in word MD50 to a 32-bit floating-point number in word MD54 and then divide the result by 100. Store the final result in word MD58.

Solution: The STL program to perform the math operation is as follows:

STL Instruction	Description of Instruction
L MD50	Load value in MD50 into accumulator 1.
L MD54	Load value in MD54 into accumulator 1.
CPU transfers MD50 value to accumulator 2.	
+R	Add MD50 to MD54.
L 1.000E+02	Load 100 into accumulator 1.
/R	Divide 100 into sum of MD50 and MD54.
T MD58	Transfer result to MD58.

Table 8-6. **Types of STL Program Language Comparisons**

Type of Comparisons	Relational Operator
Accumulator 2 numeric value is equal to accumulator 1 numeric value	==
Accumulator 2 numeric value is not equal to accumulator 1 numeric value	<>
Accumulator 2 numeric value is greater than accumulator 1 numeric value	>
Accumulator 2 numeric value is less than accumulator 1 numeric value	<
Accumulator 2 numeric value is greater than or equal to accumulator 1 numeric value	>=
Accumulator 2 numeric value is less than or equal to accumulator 1 numeric value	<=

Compare Single Integer

The Compare Single Integer (I) instruction compares the two 16-bit whole numbers loaded in accumulators 1 and 2. This instruction compares the numeric value according to the type of comparison that is selected from the list of relational operators in Table 8-6.

If the comparison is TRUE, the result-of-logic-operation (RLO) bit in the status word is set to one (1). If the comparison is FALSE, the RLO bit in the status word is set to zero. The status of the RLO bit can be used in other parts of your STL program. The CPU executes the compare instruction without regard to the result of a logic operation.

The STL program shown in Figure 8-22 uses three different types of integer compare instructions to illustrate their operation. The first instruction in the program, L 5, loads the number 5 into accumulator 1. The second instruction, L MW10, loads the integer value in memory word MW10 into the Low Word of accumulator 1, and the number 5 is shifted to the Low Word in accumulator 2. The third line in the program contains a Compare Equal ("==I") instruction. It compares the number 5 in the Low Word of accumulator 2 to the integer value in the Low Word of accumulator 1 to see if they are equal. If the value in word MW10 is equal to 5, then the output bit Q124.0 is set to one (1) by the assign instruction =Q124.0 in the fourth line of the program.

The fifth line contains a Compare Greater Than (">I") instruction. It compares the number 5 in the Low Word of accumulator 2 to see if it is greater than the value in the Low Word of accumulator 1. If the number 5 is greater than the integer value in word MW10, the output bit Q124.1 is set to one by the assign instruction =Q124.1 in the sixth line of the program. The seventh line contains a compare-less-than ("<I") instruction. It compares the number 5 in the Low Word of accumulator 2 to see if it is less than the value in the Low Word of accumulator 1. If the number 5 is less than the integer value in word MW10, then the output bit Q124.2 is set to one by the assign instruction =Q124.2 in the last line of the program.

STL Instructions	Description
L 5	Load the number 5 into accumualtor 1
L MW10	Load the integer value in MW10 into accumualtor 1
==I	Determine if the integer value of MW10 is equal to 5.
= Q124.0	Set Q124.0 to 1, if MW10 = 5
>I	Determine if MW10 is greater than 5
= Q124.1	Set Q124.1 to 1, if MW10 > 5
<I	Determine if MW10 is less than 5
= Q124.2	Set Q124.2 to 1, if MW10 < 5

Figure 8-22. Example STL program using Compare Single Integer instructions.

Compare Double Integer

The *Compare Double Integer* instruction compares the two 32-bit whole numbers loaded in accumulators 1 and 2. The instruction compares the numeric value according to the type of comparison that is selected from the list of relational operators in Table 8-6.

If the comparison is TRUE, the result-of-logic-operation (RLO) bit in the status word is set to one (1). If the comparison is FALSE, the RLO bit in the status word is set to zero. The status of the RLO bit can be used in other parts of the STL program. The CPU executes the compare instructions without regard to the result of a logic operation.

The STL program shown in Figure 8-23 uses three different double integer compare instructions to illustrate their operation. The first instruction in the program, L MD2, loads the integer value in double word MD2 into accumulator 1. The second instruction, L MD10, loads the integer value in memory word MD10 into accumulator 1, and the old value is shifted into accumulator 2. The third line in the program contains a Compare Equal ("==D") instruction. It compares MD2 in accumulator 2 to the double integer value (MD10) in accumulator 1 to see if they are equal. If values are equal, then the output bit Q124.3 is set to one by the assign instruction =Q124.3 in the fourth line of the program.

The fifth line of the program contains a double integer Compare Greater Than (">D") instruction. It compares the integer value from double word MD2 in accumulator 2 to see if it is greater than the integer value in accumulator 1 from double word MD10. If the integer value from double

STL Instructions	Description
L MD2	Load the value in double word MD2 into accumualtor 1
L MD10	Load the integer value in MW10 into accumualtor 1
==D	Determine if the values in MD2 and MD10 are equal
= Q124.3	Set Q124.3 to 1, if MD2=MD10
>D	Determine if MD10 is greater than MD2
= Q124.4	Set Q124.4 to 1, if MD10 >MD2
<D	Determine if MD10 is less than MD2
= Q124.5	Set Q124.5 to 1, if MD10 < MD2

Figure 8-23. Example STL program using Compare Double Integer instructions.

word MD2 is greater than the integer value from word MD10, then the output bit Q124.4 is set to logic 1 by the assign instruction =Q124.4 in the sixth line of the program. The seventh line contains a Compare Less Than ("<D") instruction. It compares the integer value from double word MD2 in accumulator 2 to test if it is less than the integer value from word MD10 in accumulator 1. If the integer value in word MD2 is less than the integer value in word MD10, the output bit Q124.5 is set to TRUE by the assign instruction =Q124.5 in the last line of the program.

Compare Real

The *Compare Real* instruction compares the two 32-bit real numbers loaded in accumulators 1 and 2. This instruction compares the numeric values according to the type of comparison that is selected from the list of relational operators in Table 8-6. If the comparison is TRUE, the result-of-logic-operation (RLO) bit in the status word is set to one (1). If the comparison is FALSE, the RLO bit in the status word is set to zero. The status of the RLO bit can be used in other parts of your STL program. The CPU executes the compare instruction without regard to the result of a logic operation.

The STL program shown in Figure 8-24 uses a Less Than and a Greater Than Compare instruction to illustrate their operation on real numbers. The first instruction in the program, L MD10, loads the value in double word MD10 into accumulator 1. The second instruction, L +1.75E01, loads the number 17.5 into the accumulator 1, and the MD10 value is shifted into accumulator 2. The third line in the program contains a Compare Greater Than Real (">R") instruction. It compares the value from double word MD10 in accumulator 2 to see if it is greater than the value of 17.5 in accumulator 1. If the number 17.5 is greater than the integer value from word MD10, then the output bit Q124.6 is set to logic 1 by the assign instruction =Q124.6 in the fourth line of the program. The fifth line contains a Compare Less Than Real ("<R") instruction. It compares the value from double word MD10 in accumulator 2 to test if it is less than the real number 17.5 in accumulator 1. If the integer value in word MD10 is less than 17.5, then the output bit Q124.7 is set to TRUE by the assign instruction =Q124.7 in the last line of the program.

STL Instructions	Description
L MD10	Load the value in double word MD10 into accumualtor 1
L 1.75E+01	Load the number 17.5 into accumualtor 1
>R	Determine if the value in MD10 is greater than 17.5
= Q124.6	Set Q124.6 to true, if MD10 > 17.5
<R	Determine if value in MD10 is less than 17.5
= Q124.4	Set Q124.7 to true, if MD10 < 17.5

Figure 8-24. Example of an STL program using a Compare Real instruction.

EXERCISES

8.1   List the five PLC programming languages defined in the IEC 61131-3 standard.

8.2   What are the two text-based PLC languages? Provide a brief description of each.

8.3   What are the three graphical PLC languages? Provide a brief description of each.

8.4   Describe the "initial" step instruction in a sequential function chart (SFC) program?

8.5   Explain the purpose of a "transition" instruction in a SFC program.

8.6   List the programming tips that should be followed when using inactive separators to improve the readability of a structured text (ST) program.

8.7   What are the two basic structures of a STL instruction statement?

8.8   Write an STL program to turn on a pump starter that is connected to output point Q124.0, if control valve FV-1 is opened (input bit I124.0 =1) and the process tank level is high (input bit I124.2 = 1).

8.9   List the five timer instructions available in the Siemens S7 STL programming instruction set.

8.10   Write a STL program for a Siemens Simatic S7-300 PLC that causes it to open a fill valve on a process tank so an ingredient can be added to the tank for thirty seconds. Assume that the fill valve is wired to PLC output point Q124.4. Also assume that an operator uses a momentary normally open pushbutton connected to PLC input point I124.6 to open the fill valve. Use timer T3 and input bit I124.2 to reset the timer in the program.

8.11   Write a STL program for a Siemens Simatic S7-300 PLC that causes it turn to off a conveyor belt on a production line after ten parts have been produced. Assume the following: (1) output bit Q124.5 = 0, turns off the conveyor belt; (2) input I124.6 changes from 0 to 1 and then back to 0 each time a production part is rejected; (3) input I124.7 changes from 0 to 1 and then back to 0 each time a new part is produced; (4) a normally open (NO) pushbutton connected to input I124.2 is used to set the production count to ten, and (5) a NO pushbutton connected to input I124.3 is used to reset the counter to zero and to stop the conveyor belt. Use counter number 4 in the STL program.

8.12   Write a Siemens Simatic S7 STL program to add the integer in word MW60 to the integer data in word MW62 and then divide the result by ten. Store the result in word MW64.

8.13   Write a Siemens Simatic S7 STL program to subtract a 32-bit floating-point number in word MD70 from a 32-bit floating-point number in word MD74 and then divide the result by ten. Store the final result in word MD78.

BIBLIOGRAPHY

1.   Christensen, J. H. "Programmable Controller Users and Makers to Go Global with IEC1131-3," *I&CS Magazine* (October 1993).

2.   Gilbert, R. A., and J. A. Llewellyn. *Programmable Controllers: Practices and Concepts* (Industrial Training Corporation, 1985).

3.   Lloyd, M. "Grafcet Graphical Functional Chart Programming for Programmable Controllers," *Measurement & Control Magazine* (September 1987).

4.   Siemens AG. *Simatic S7 Statement List (STL) for S7-300 and S7-400 Programming Reference Manual* (Siemens AG, October 1998).

9

Function Block
Diagram
Programming

Introduction

The function block diagram (FBD) is a graphical programming language that uses logic blocks similar to those used in Boolean algebra to represent basic logic. It also uses more complex function blocks to perform operations such as timing, counting, math, loading data, transferring data, and comparing data. The programmer is able to build complex control schemes by using functions from the FBD library and then interconnecting them in a graphical diagram area.

In this chapter, we will discuss the basics of FBD programming as well as some typical applications. We will also explore the FBD programming language used in Siemens S7-300 and S7-400 programmable controllers. Most of the other PLC manufacturers use a similar version of the standard FBD language. However, since we will only cover the basics of Siemens FBD programming, the Siemens reference manuals should be consulted for more detailed information on the use of their software.

Elements and Box Structure

FBD instructions consist of elements and boxes that are connected graphically to form networks. An FBD box describes a relationship or function between input and output variables. A function is a set of elementary function blocks, as shown in Figure 9-1. Input and output variables are connected to blocks by connection lines. The programmer can build an entire FBD program with standard elementary function blocks from the FBD library and also develop custom function blocks.

Each elementary function block has a given number of input connection points and output connection points. For example, the Boolean AND function block shown in Figure 9-1 has two inputs and only one output. The number of inputs can be increased to a higher value but the output is limited to one for this AND function box instruction. However, the output can be branched off to any number of other function blocks. The inputs are connected on its left border; the outputs are connected on its right border. An elementary box performs a single function between its inputs and its outputs. For example, the elementary function block shown in Figure 9-1 performs the Boolean AND operation on its two inputs and produces a result at the output. The name of the function to be performed by the block is written in its symbol. In the case of the AND function the symbol is &.

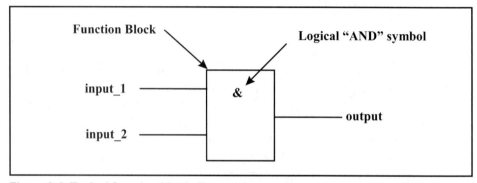

Figure 9-1. Typical function block diagram instruction.

Elements and boxes that are connected graphically to form logic networks can be classified into the following types:

- Instruction as elements

- Instruction as a box with an address

- Instruction as a box with an address and value

- Instruction as a box with parameters

Table 9-1 provides examples of the four types of elements and boxes used in FBD programming.

Table 9-1. List of Typical FBD Instruction Types

Type	Example	Element or Box
1. Instruction as elements	Negate binary input	<input> ———o|
2. Instruction as a box with an address	Assign	<address> ——[=]
3. Instruction as a box with an address and value	Retentive on-delay timer	<address> —[SS] <Time Value>—[TV]
4. Instruction as a box with parameters	Divide real	DIV_R —EN —IN1 OUT— —IN2 ENO—

Bit Logic Instructions

Boolean bit logic instructions are the most basic type of FBD instructions. They perform logic operations on single bits in PLC memory. The basic bit logic instructions are AND (A), OR (O), and EXCLUSIVE OR (XOR). These instruction boxes check the logic state of an input bit address to establish whether the bit is activated, 1, or not activated, 0.

AND Logic Operation

In an AND logic operation, the logic state (1 or 0) of two or more specified input addresses is tested. If all inputs are 1, the output of the AND (&) function block is set to one (1). If any input is zero, the output of the AND function block is set to zero. Figure 9-2 shows an example of a two-input AND function block instruction in which the output from the function box is assigned to output bit location Q124.0. If the state of the two inputs I124.0 and I124.1 is set to one, then the bit in the assigned output Q124.0 is set to one. If the AND instruction is the first instruction in a string of logic operations, it saves the result of its logic state test in the Result of the Logic Operation (RLO) bit of the processor's status word.

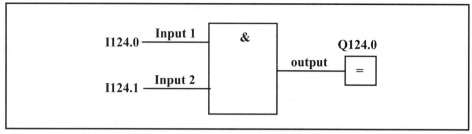

Figure 9-2. Example of an AND function block instruction.

The two input bits (I124 and I124.1) and the output bit are bit addresses for a Siemens Simatic programmable controller, model number CPU314IFM. This is a compact PLC with twenty digital inputs, sixteen digital outputs, four analog inputs, and one analog output. This compact PLC is designed for small control applications. Figure 9-3 is an input/output (I/O) wiring diagram for the Simatic CPU314IFM. It shows the eight digital inputs, with addresses of I124.0 through I124.7, and eight digital outputs, with addresses Q124.0 through Q124.7.

The wiring diagram also shows eight test switches and pushbuttons connected to the eight-digital inputs. The sixteen I/O points will be used in the example and application programs in this chapter. The input switches and pushbuttons can be used to test the control program. Each of the sixteen digital I/O points has LED (light-emitting diode) lights to indicate its status. If the LED is ON, the I/O bit is set to logic 1. If one of the normally opened (NO) hand switches (HS-1 though HS-4) is closed, +24 volts direct current (vdc) is applied to the input point, the associated logic bit (I124.0 though I124.3) will be set to one (1), and the LED status light will be turned on.

Pushbuttons PB1 and PB2 are normally closed (NC). As a result, when the pushbuttons are not depressed +24 volts DC are applied to inputs I124.4 and I124.5, the LED for inputs I124.4 and I124.5 will be turned on, and the logic bits will be set to one. If the NC pushbuttons are pressed, the DC voltage is removed from the input points, and the input bits will be set to zero. The LEDs on the digital output points (Q124.0 through Q124.7) are turned on if the control program sets the associated bits to logic 1.

You can use the PLC shown in Figure 9-3 with its test switches and pushbuttons to test the operation of a new PLC program. The input switches and pushbuttons can be operated (opened or closed) and the LED lights on the output points can be monitored to verify that a control program is operating properly before it is installed in an industrial application.

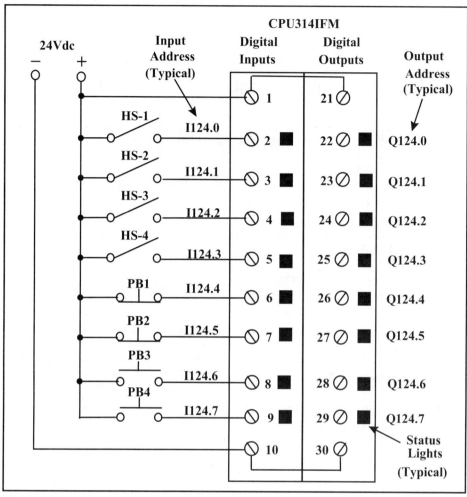

Figure 9-3. Wiring diagram for the sixteen digital I/Os on a Siemens CPU314IFM.

OR Logic Operation

In an OR logic operation, the logic state of two or more specified bit addresses is tested. If the logic state of one of the input bit addresses is one (1), the output of the OR function block is one. If the logic state of all the input bit addresses is zero, the output is set to zero. Figure 9-4 shows a typical example of an OR function block instruction with inputs at I124.2 and I124.3 and the output of the function box assigned to output bit address Q124.1. The output is set to one if the logic state of input I124.2 or input I124.3 is set to one.

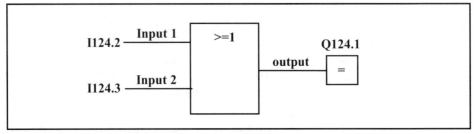

Figure 9-4. Example of OR function block instruction.

An example will help to illustrate FBD programming using AND and OR instructions.

EXAMPLE 9-1

Problem: Write an FBD program to start and stop a pump. In this application, the normally open (NO) contacts of a start pushbutton are wired to input bit address I124.6. Moreover, the normally closed (NC) contacts of a stop pushbutton are connected to input bit address I124.4 of a Siemens Compact PLC model CPU314IFM (see Figure 9-3). The pump starter relay is connected to PLC output address Q124.0, and the auxiliary (aux.) motor start contacts (NO) are connected to PLC input bit address I124.0.

Solution: The application can be implemented using the program shown in Figure 9-5.

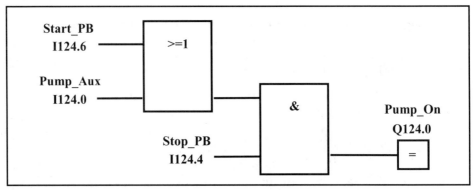

Figure 9-5. Example of pump start/stop FBD program.

When the NO start pushbutton is depressed, input I:1/1 is TRUE, so the output of the OR function block instruction is one (1). Moreover, since the NC stop pushbutton is not depressed, input I:1/0 is also TRUE, so both inputs to the AND box are TRUE. Since both inputs to the AND instruction block are one, the output of the AND block is set to one. This sets the bit address Q124.0 to one. Since this output bit is wired to the pump start relay, the pump start relay is turned on, which causes the pump auxiliary (aux.) contacts to close. This, in turn, sets the *Pump_Aux* input bit I124 to keep the output of the OR box at one even after the start pushbutton is released. If the stop pushbutton is depressed, the *Stop_PB bit, I124.4:* is set to zero and the *Pump_On* output bit is deenergized or set to zero. As a result, the pump will be turned off, and the auxiliary contacts on the pump starter will open and set the *Pump_Aux* input bit I124.0 to zero.

EXCLUSIVE OR Logic Operation

In an EXCLUSIVE OR logic operation, the logic state of two or more specified bit addresses is tested. If the logic state of one and only one of the input bit addresses is one (1), the output of the OR function block is one. If the logic state of all the input bit addresses is zero or one, the output is set to zero. Figure 9-6 shows a typical example of an EXCLUSIVE OR function block instruction with inputs at I124.4 and I124.5 and the output of the function box assigned to output bit address Q124.2. The output is set to one if the logic state of input I124.2 or input I124.3 is set to one, but the output is set to zero if both inputs are one or zero.

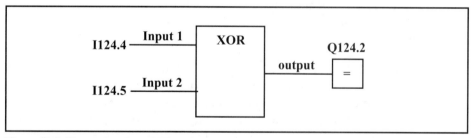

Figure 9-6. Example of an EXCLUSIVE OR instruction.

Negate Binary Input

The **Negate Binary Input** instruction is represented by a small circle at the input to a function box. This FBD element-type instruction inverts or negates a bit that is present at the input to the element. For example, in Figure 9-7, if the bit at input address I124.1 is logic 0, it is negated or inverted to a logic 1 by the negate element before it is operated upon by the AND instruction. On the other hand, if the same input (I124.1) is logic 1, it is negated to a logic 0 before the AND function block uses its value. The truth table in Figure 9.7 shows the effect of the **Negate Binary Input** element on input bit I124.1 and on the operation of the AND instruction. The first column of the truth table lists the three possible values for I124.1.

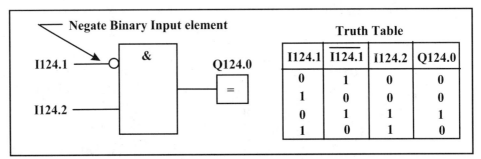

Figure 9-7. Example of a Negate Binary Input element instruction.

The second column lists the negated value of I124.1. The third column lists the logic states of the second input bit, I124.2. The last column provides the AND logic result for the second and third columns.

Assign

The **Assign** instruction is represented by an equal sign (=) in a function box, as shown in Figure 9-7. This FBD logic operation assigns the logic state of the bit at the input to the box to the address listed above the function box. For example, in Figure 9-7 the result of the AND operation is assigned to the address Q124.0. The bit address types that can be used with the **Assign** instruction are I (input), Q (output), M (memory), D (data), and L (local data).

Set Output

The **Set Output** instruction is represented by the letter *S* in a function block. The **Set Output** instruction is only executed when the input to the box is logic 1. If the input bit is one (1), this instruction sets the specified bit address to one. If the input to the block is zero, the instruction does not affect the specified bit address, and the logic state remains unchanged. The memory address types that can be used with the **Set Output** instruction are the I (input), Q (output), M (memory), T (timer), C (counter), D (data), and L (local data) bit addresses.

Reset Output

The **Reset Output** instruction is represented by the letter *R* in a function block. The **Reset Output** instruction is only executed when the input to the box is logic 1. If the input bit is one (1), this instruction resets the specified bit address to zero. If the input to the reset instruction block is zero, the instruction does not affect the specified bit address and the logic state remains unchanged. The memory address types that can be used with the **Set Output** instruction are the I (input), Q (output), M (memory), T (timer), C (counter), D (data), and L (local data) bit addresses.

Application Using Bit Instructions

To illustrate the use of the basic logic bit instructions, we will now discuss the control of the conveyor belt shown in Figure 9-8. In this application, there are two sets of start and stop pushbuttons, one at the beginning of the conveyor and one at the end of the belt. These pushbuttons (PB1 through PB4) are wired to input points on a Siemens Simatic CPU314IFM programmable controller, as shown in Figure 9-9. The two start buttons use normally open (NO) contacts, and they are labeled PB1 and PB3. The two stop buttons use normally closed (NC) contacts, and they are labeled PB2 and PB4. The control program allows the operator to start and stop the conveyor belt from either end. There is a position detection switch (ZS1) at the end of the conveyor to sense when a production part reaches the end of the conveyor. This input signal is used by the PLC control program to automatically stop the conveyor when a part reaches the end of the conveyor.

The square box with a diamond inside is the ISA standard symbol used to indicate a PLC input or output point on process and mechanical diagrams. The dashed line with circles between each dash is the ISA standard symbol for a computer software link.

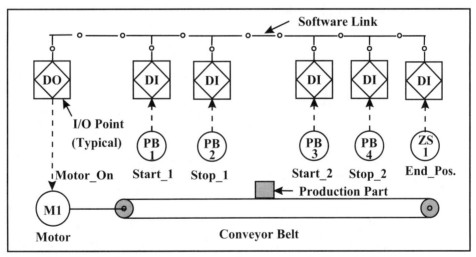

Figure 9-8. Mechanical process diagram for conveyor belt application.

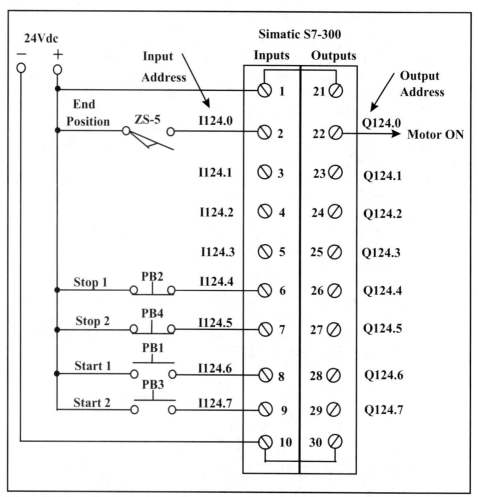

Figure 9-9. Input/output wiring diagram for conveyor belt application.

The function block program for the conveyor belt application is given in Figure 9-10. In this program, the symbolic addresses for the I/O points are used instead of the absolute addresses. Table 9-2 provides the cross-reference between the absolute I/O addresses and the symbolic addresses. Most PLC programming software packages make it possible to symbolically represent absolute I/O to help you program and test control programs. It is much easier to understand and troubleshoot a control program that uses symbolic I/O memory addresses than one that uses absolute addresses.

The first column of Table 9-2 contains a list of the system instruments and components, and the second column gives the absolute PLC I/O address for each component. The third column provides the symbol that is used to replace the I/O address, and the last column is the symbol table, which is used by the PLC program to make the program easier to understand.

Table 9-2. Elements for Symbolic Programming of Conveyor Belt System

System Component	Absolute Address	Symbol	Symbol Table
Stop 1 NC PB	I124.4	Stop_1	I124.4 Stop_1
Stop 2 NC PB	I124.5	Stop_2	I124.5 Stop_2
Start 1 NO PB	I124.6	Start_1	I124.6 Start_1
Start 2 NO PB	I124.7	Start_2	I124.7 Start_2
Pos. Sw. ZS1	I124.0	End_Pos	I124.0 End_Pos
Motor Starter	Q124.0	Motor_On	Q124.0 Motor_On

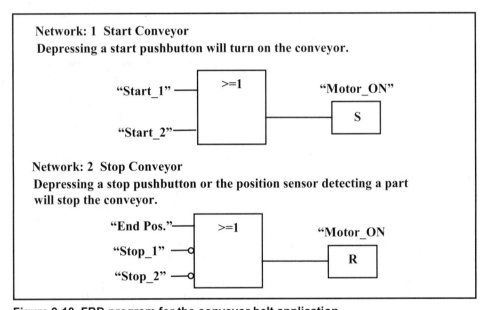

Figure 9-10. FBD program for the conveyor belt application.

Timer Instructions

Software timers provide the same functions as hardware timers in process control applications. A typical application for a timer instruction is to delay an operation for a fixed time interval. For example, the starting of a pump might be delayed for several seconds until a valve on the discharge line of the pump is completely opened so high pressure cannot develop on the discharge of the pump. Another example would be to have a timer determine the amount of time it takes a vessel to fill with fluid and then display that time to the plant operator on a graphics display.

Timer Memory Word Structure

Timer instructions in a Siemens S7 PLC require a sixteen-bit word in memory, as shown in Figure 9-11. Bits 0 through 11 of the timer word contain the time value in binary-coded decimal (BCD). Bits 12 and 13 of the timer contain the time base in binary code. The left-most two bits (bits 14 and 15) in the timer word are irrelevant and ignored when the timer is started. Figure 9-11 shows the timer word loaded with a timer value of 349 and with a time base of one second. The twelve bits in cell locations 0 through 11 are divided into three groups of four bits each to produce a three-digit BCD number.

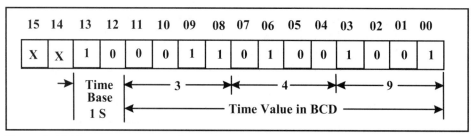

Figure 9-11. Timer memory word structure.

Timer Function Block Configuration

The programmer selects a time value (TV) or timing interval for a timer by using the following configuration syntax: S5T#aH_bbM_ccS_dddMS, where *aa* is time value in hours, *bb* is a time value in minutes, *cc* is a time value in seconds, and *ddd* is a time value in milliseconds. For example, to program a timer for ten seconds, the programmer would type in "S5T#10S" when configuring the TV input to the timer function block. If a time value of one hour and thirty minutes were required, the programmer would input "S5T#1H30M" into the timer function block TV input during configuration.

There are four time bases available: 10 ms, 100 ms, 1 second, and 10 seconds. Table 9-3 lists the time bases and their corresponding binary code. The time base is selected automatically by the PLC when the function block is configured for a time value.

Table 9-3. Time Base and Its Binary Code

Time Base	Binary Code for Time Base
10 ms	00
100 ms	01
1 second	10
10 seconds	11

Pulse S5 Timer

The *Pulse S5 Timer* starts timing when there is a change in the logic input state from zero to one (1) at the start (S) input. A change in logic state is always required to start a Siemens S5 timer. The *Pulse S5 Timer* will continue to run for the time interval specified at the TV input until the programmed time elapses, as long as the logic state of the S input is one and the reset (R) is not set to logic 1 during the timing interval. If the S input changes from one to zero before the timing interval is complete, the timer is stopped and the output Q goes to zero.

If the signal at the reset input changes from zero to one while a timer is running, the timer is stopped and the time is reset to zero. A logic state of one at the R input of the timer has no effect if the timer is not running. Figure 9-12 illustrates the operation of a Pulse S5 Timer instruction. In this application, input bit I124.1 is connected to the set input, and input bit I124.6 is connected to the reset input. The timing diagram shows the operation of the timer for various states of the R and S inputs. The timer output is assigned to output bit Q124.0, and the programmed time (t) interval is ten seconds.

The timer has two word outputs, BI and BCD, to indicate the time remaining. The time value in the BI output word is in binary format, and the value in the BCD output word is in binary-coded decimal format. The programmer configures the timer instruction for the memory word address for BI and BCD words if this timer information is needed in an

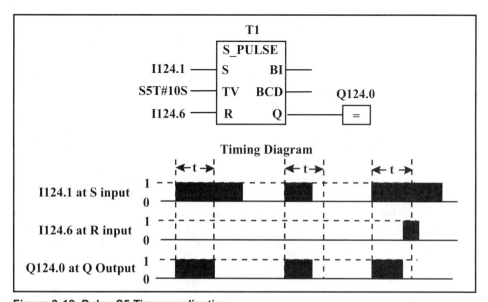

Figure 9-12. Pulse S5 Timer application.

application. In the example shown in Figure 9-12, the BI and BCD output words are not assigned to any particular memory address.

Extended Pulse S5 Timer

The *Extended Pulse S5 Timer* starts timing when there is a change in the logic input state from zero to one at the start (S) input. A change in logic state at the start input is always required to start a Siemens S7 timer. The Extended Pulse S5 Timer will continue to run for the time interval specified at the TV word input until the programmed time elapses. This is the case even if the logic state of the S input is not one (1) and the reset (R) input is not changed from zero to one during the timing interval. However, if the S input changes from zero to one before the timing interval is complete, the timer will still continue to run to the end of the programmed time value.

If the signal at the reset input changes from zero to one while a timer is running, the timer is stopped and the time is reset to zero. A logic state of one at the R input of the timer will prevent the timer from starting on a zero to one transition at the start input. The Extended Pulse timer has two word outputs, BI and BCD, to indicate the time remaining in a run cycle. The time value in the BI output word is in binary format, and the value in the BCD output word is in binary-coded decimal format. The programmer configures the timer instruction for the memory word address for BI and BCD words if this timer information is needed in an application.

Figure 9-13 illustrates the operation of an **Extended Pulse S5 Timer** instruction. In this application, input bit I124.1 is connected to the set input, and input bit I124.2 is connected to the reset input. The programmed time value is ten seconds (i.e., TV = S5T10s). The timing diagram in Figure 9-13 shows the operation of the timer for various states of the R and S inputs. The timer output is assigned to output bit Q124.0, and the programmed time value of ten seconds is represented by the letter *t* on the timing diagram.

As the timing diagram in Figure 9-13 shows, the Extended Pulse Timer is more complex than a pulse timer. There are four different conditions shown on the timing diagram. In the first condition, the start input goes from zero to one at time t_0 and remains at one for more than ten seconds. The timer runs for the set time value of ten seconds, and the Q output is one for ten seconds. In the second condition, the start signal goes from logic 0 to logic 1 at time t_2, but the start input returns to zero after a short time. However, the timer continues to run, and the Q output is logic 1 while the timer is running. Then, after ten seconds at time t_3, the timer stops and the Q output returns to logic 0. The third condition starts at t_4 when the start input goes to logic 1 for a few seconds and then returns to

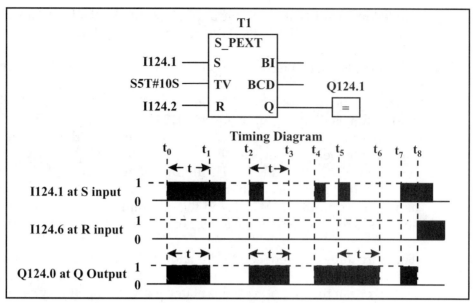

Figure 9-13. Extended Pulse S5 timer application.

logic 0. The timer runs normally, but a few seconds later a second start signal is applied. The timer continues to run, but the time interval is *extended* by an additional ten seconds until time t_6, hence the name *Extended* Pulse Timer. The final condition shown in Figure 9-13 is a standard timer reset operation. The timer is started at time t_7, but several seconds later, at time t_8, the reset input at I124.2 goes from zero to one, and the timer is reset and the Q output goes to zero.

On-Delay S5 Timer

The **On-Delay S5 Timer** instruction starts timing when there is a change in the logic input state from zero to one at the start (S) input. This type of timer counts down the time interval specified at the TV input as long as the S input is not logic 1. The Q output is changed from zero to one only after the programmed time elapses. In other words, the output is turned ON only after a *delay* in the time interval programmed in the timer instruction. Hence, the timer is named *On-Delay*.

If the signal at the reset input changes from zero to one while a timer is running, the timer is stopped and the time is reset to zero. The On-Delay S5 Timer has two word outputs, BI and BCD, to indicate the time remaining in a timing interval. The time value in the BI output word is in binary format, and the value in the BCD output word is in binary-coded decimal format. The programmer configures the timer instruction for the memory word address for BI and BCD words if this timer information is needed in an application. In the application of On-Delay S5 Timer shown in Figure 9-14, the BI and BCD outputs are assigned to memory word 10.

In this application, a normally open "Run" switch is wired to input address I124.3 on a Simatic S7 PLC. This input address is connected via programming to the set input of On-Delay Timer T3. A normally closed pushbutton is wired to input address I124.4 on a Simatic S7 PLC. This input address point is connected via programming to the negated reset input of On-Delay Timer. The timer is programmed for a thirty-second time value (i.e., TV = S5T30s). The timing diagram shows the operation of the timer for various states of the R and S inputs. The timer output is assigned to output bit Q124.2, and the programmed time value of 30 seconds is represented by the letter t on the timing diagram.

There are three different conditions shown on the timing diagram. In the first condition, the "Run" switch is closed and the start input goes from zero to one at time t_0 and remains at one for more than thirty seconds. The timer runs for the set time value of thirty seconds and then sets the Q output to one at time t_1. This output is connected to output Q124.2 on the PLC, which in turn is wired to the pump start relay shown in Figure 9-14. The motor start relay is energized and remains on until time t_3, when the "Run" switch is opened by an operator, and the start input to the timer goes from one to zero.

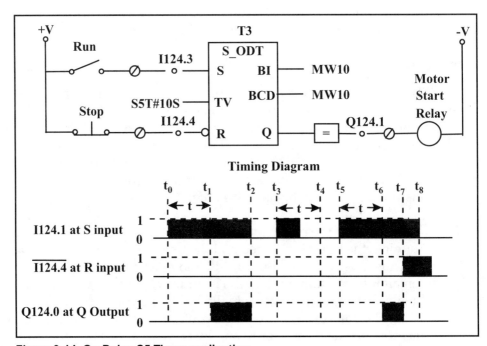

Figure 9-14. On-Delay S5 Timer application.

In the second condition, the start signal goes from logic 0 to logic 1 at time t_4, but the start input returns to zero at time t_5 before thirty seconds have elapsed, and the Q output remains at zero.

The final condition shown in Figure 9-14 is a reset of the timer with the start input still at logic 1 and at a point after the timer output has been activated. The timer is started at time t_6, and the output Q is activated after a delay of thirty seconds at time t_7. At time t_8, the stop pushbutton is depressed, so input I124.4 changes from one to zero, and the negated reset input goes from zero to one, resetting the timer. The Q output goes to zero and turns off the motor start relay.

Retentive On-delay S5 Timer

The **Retentive On-delay S5 Timer** instruction starts timing when there is a change in the logic input state from zero to one at the start (S) input. This timer continues to run for the time specified at the time value (TV) input even if the signal state at input S changes from one to zero before the time has expired. In other words, the Q output is retained at one after the programmed time elapses even if the S input is not kept positive.

If the signal at the reset input changes from zero to one while a timer is running, the timer is reset to zero. The Retentive On-delay S5 Timer has two word outputs, BI and BCD, to indicate the time remaining in a timing interval. The time value in the BI output word is in binary format, and the value in the BCD output word is in binary-coded decimal format. The programmer configures the timer instruction for the memory word address for BI and BCD words if this timer information is needed in another part of an application program. In the **Retentive On-delay S5 Timer** instruction shown in Figure 9-15, the BI and BCD outputs are assigned to memory word 10.

In this application, a normally open "start" switch is wired to input address I124.0 on a Simatic S7 PLC. This input address is connected via internal software to the set input of timer instruction. A normally open "reset" switch is wired to input address I124.1. This input address point is connected via internal software to the reset input of timer instruction. The timer is programmed for a thirty-second time value (i.e., TV = S5T30s).

The timing diagram shows the operation of the timer for various states of the R and S inputs. The timer output is assigned to output bit Q124.2, and the programmed time value of thirty seconds is represented by the letter t on the timing diagram.

There are three different operations shown on the timing diagram. In the first operation, the "start" switch is closed, and the start input goes from

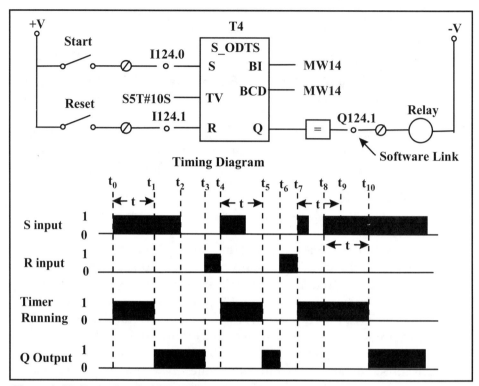

Figure 9-15. Retentive On-delay S5 timer application.

zero to one at time t_0 and remains at one for more than thirty seconds. The timer runs for the set time value of thirty seconds and then sets the Q output to one at time t_1. This output is connected to output Q124.2 on the PLC, which in turn is wired to a relay, as shown in Figure 9-15. The relay is energized and remains on until time t_3, when the "reset" switch is closed by an operator, the reset input to the timer instruction goes from zero to one, and the timer Q output is set to zero.

In the second condition, the start signal goes from logic 0 to logic 1 at time t_4 with the reset input set to logic 0. Before the time interval has reached thirty seconds, the start signal is returned to zero, but the timer continues to run until time t_5, when the timer output Q is set to one. At time t_6, the timer is reset, and the output Q is deactivated.

In the final operation, at time t_7 the start input is changed from zero to one and the timer instruction starts. However, before the programmed time has elapsed, the start input is returned to zero and then back to one. This restarts the timer at time t_8. The timer runs until time t_{10} and then sets the Q output to one. It will remain at one until the timer is reset.

Off-Delay S5 Timer

The **Off-Delay S5 Timer** instruction starts timing when there is a falling edge (change in signal state from zero to one) at the start (S) input. The output Q is logic 1 when the logic state of the S input is one (1) or when the timer is running.

If the signal at the reset (R) input changes from zero to one, the timer is reset to zero and stops running. The Off-Delay S5 Timer has two word outputs, BI and BCD, to indicate the time remaining in a timing interval. The time value in the BI output word is in binary format, and the value in the BCD output word is in binary-coded decimal format. In the Off-Delay S5 Timer instruction application shown in Figure 9-16, the BI and BCD outputs are assigned to memory word 10.

In this application, a normally open "start" switch is wired to input address I124.0 on a Simatic S7 PLC. This input address is connected via internal software to the set input of timer instruction. A normally open "reset" switch is wired to input address I124.1. This input address point is connected via internal software to the reset input of the timer instruction. The timer is programmed for a thirty-second time value (i.e., TV = S5T30s).

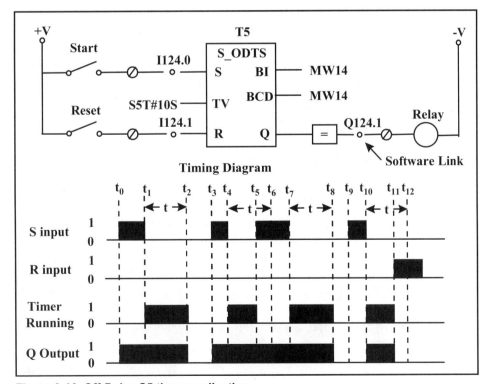

Figure 9-16. Off-Delay S5 timer application.

The timing diagram in Figure 9-16 shows the operation of the timer for various states of the R and S inputs. The timer output is assigned to output bit Q124.2, and the programmed time value of thirty seconds is represented by the letter t on the timing diagram.

There are three different operations shown on the timing diagram. In the first operation, the "start" switch is closed, and the start input goes from zero to one at time t_0, the start switch is opened at t_1, and the timer runs for thirty seconds until time t_2. The Q output is set to one from time t_0 to t_2. The output of the timer is connected to output bit Q124.2 on the PLC, which in turn is wired to a relay, as shown in Figure 9-16.

In the second condition, the start signal is changed from zero to one and one to zero several times. Starting at time t_3, the "start" switch is closed by an operator, which changes the start input signal from zero to one. This sets output Q to one. At time t_4, the start signal goes from logic 1 to logic 0. This starts the timer so the output remains at one. Before the time interval has reached thirty seconds, the start signal is again returned to one at time t_5, so the timer stops, but the timer output Q remains set to one. At time t_7, the start input goes from one to zero and the timer runs again. This keeps the output Q active until time t_8, when the timer completes its programmed time interval.

In the final operation, the timer is reset during a timing operation as follows: at time t_9, the start input is changed from zero to one, and the timer output Q is activated. At time t_{10}, the start input is changed from one to zero, and the timer instruction starts. However, before the programmed time interval has elapsed, the reset input is changed from zero to one. This stops the timer at time t_{11} and sets the Q output to zero. It will remain at zero until the timer is started or the start input is set to one.

Counter Instructions

Software counters provide the same types of functions as the hardware counters used in process control applications. That is, they are used to activate or deactivate a device after a selected count value (CV) has been reached. A typical application for a counter instruction is to stop a production assembly operation after a fixed number of parts have been produced. For example, a conveyor line might be stopped after a given number of parts have passed a certain point on the conveyor belt.

Counter Memory Word Structure

Counter instructions in a Siemens S7 PLC require a sixteen-bit word in memory, as shown in Figure 9-17. Bits 0 through 9 of the Siemens counter word contain the count value in binary code, as shown in the bottom of

Figure 9-17. The top memory word in Figure 9-17 shows the counter word loaded with a counter value of 349. The twelve bits in the top cell locations 0 through 11 are divided into three groups of four bits each to produce a three-digit binary-coded decimal (BCD) number for the counter.

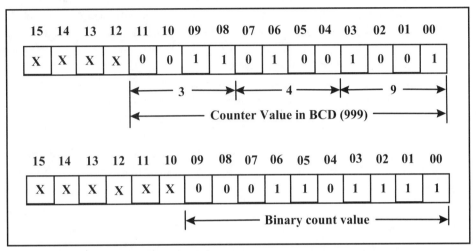

Figure 9-17. Counter memory word structure.

Counter Function Block Configuration

The programmer selects a preset value (PV) for a counter by using the following configuration syntax: C#ccc, where *ccc* is the count value (CV) in binary-code decimal (BCD) format. For example, to program a counter for a preset value of 130 counts, the programmer would type in *C#130* when configuring the PV input to a counter function block instruction. The range of preset count values is 0 to 999. The program can be configured to increment or decrement the count value within this range using the **Up/Down Counter**, **Up Counter**, and **Down Counter** instructions.

Up/Down Counter

The **Up/Down Counter** instruction counts when there is a rising edge change in the logic input state from zero to one at the CU (Count Up) input or at the CD (Count Down) input. The counter value (CV) is incremented by one if the CU input changes from zero to one (rising edge) and the value of the counter is less than 999. The counter is decremented by one if the CD input changes from zero to one and the value of the counter is greater than zero. If there is a rising edge signal at both count inputs, both operations are executed and the count remains the same. The counter is reset to a count value of zero if there is a zero to one transition at the reset (R) input.

Figure 9-18 shows a typical application that illustrates the operation of an **Up/Down Counter** instruction. There are four normally open switches connected to four input points I124.0 through I124 on a Siemens S7-300 PLC to test the operation of the Up/Down Counter. If the "Set" switch connected to input I124.2 is closed, the signal at the S input to the counter instruction is changed from zero to one. This operation sets the counter value to 44. If the "Up" switch connected to input point I124.0 is closed, the signal at the CU input to the counter instruction is changed from zero to one, and the value of counter C10 is incremented by one, except when the value of the counter is already 999. If the "Down" switch connected to input point I124.1 is closed, the signal at the CD input to the counter instruction is changed from zero to one, and the value of counter C10 is decremented by one, except when the value of the counter is already zero.

The counter status bit Q is assigned to Q124.7 in the application shown in Figure 9-18. This status bit is one when the counter value is not equal to zero.

Up Counter

The **Up Counter** instruction counts when there is a rising edge change in the logic input state from zero to one at the CU (Count Up) input. The counter value (CV) is incremented by one if the CU input changes from zero to one (rising edge) and the value of the counter is less than 999. The

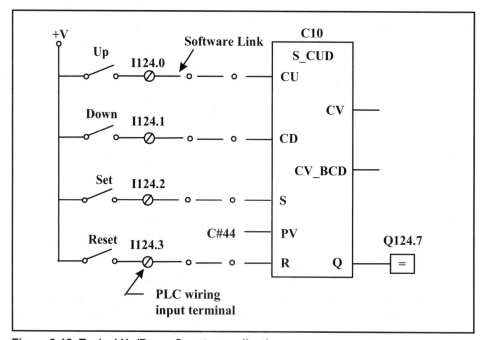

Figure 9-18. Typical Up/Down Counter application.

counter is reset to a count value of zero if there is a zero-to-one transition at the Reset (R) input.

Figure 9-19 shows a typical application illustrating the operation of an **Up Counter** instruction. There is a production part detection switch connected to input point I124.0 on a Siemens S7-300 PLC for counting the number of parts produced in a given production run. There is a "Start" switch connected to input I124.2. If this "Start" switch is closed, the signal at the S input to the counter instruction is changed from zero to one. This operation sets the counter value to 900. If the switch connected to input point I124.0 is closed, the signal at the CU input to the counter instruction is changed from zero to one, and the value of counter C11 is incremented by one, except when the value of the counter is already 999.

The counter status bit Q is assigned to Q124.7 in the application shown in Figure 9-19. This output bit is one (1), when the counter value is not equal to zero.

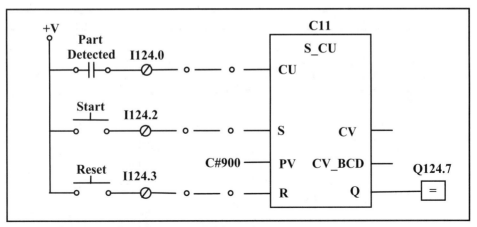

Figure 9-19. Typical Up Counter application.

The counter can be reset to zero by depressing the "Reset" switch connected to input I124.3.

Down Counter

The **Down Counter** instruction counts down when there is a rising edge change in the logic input state from zero to one at the CD (Count Down) input. The counter value (CV) is decremented by one if the CD input changes from zero to one (rising edge) and the value of the counter is greater than zero. The counter is reset to a count value of zero if there is a zero-to-one transition at the Reset (R) input.

Figure 9-20 shows a typical application illustrating the operation of a Down Counter instruction. There is a production part detection switch connected to input point I124.0 on a Siemens S7-300 PLC to count the number of parts produced in a given production run. There is a "Start" switch connected to input I124.2. When this "Start" switch is closed, the signal at the S input to the **Down Counter** instruction is changed from zero to one. This operation sets the counter value to 100. If the switch connected to input point I124.0 is closed, the signal at the CD input to the counter instruction is changed from zero to one and the value of counter C12 is decremented by one, except when the value of the counter is already zero.

The negated value of the counter output Q bit is assigned to Q124.7. This output bit is connected to a Stop Conveyor relay, as shown in Figure 9-20. The Q status bit is one (1) when the counter value is not equal to zero, so its negated bit will be one when the count value is zero. The purpose of the application is to count parts produced on a conveyor line and then stop the conveyor after one hundred parts have been produced.

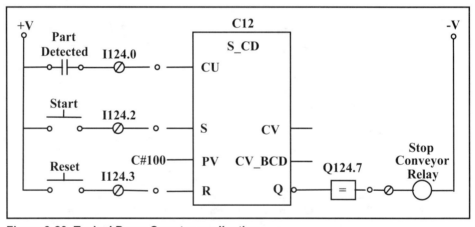

Figure 9-20. Typical Down Counter application.

The counter can be reset to zero by depressing the Reset switch connected to input I124.3.

Integer Math Instructions

In this section, we will discuss the FBD instructions used to perform mathematical operations on positive and negative whole numbers (1, 2, 3, etc.) or zero. The basic math instructions—add, subtract, multiply, and divide—for single and double integer words are listed in Table 9-4.

The integer math FBD instruction for an Add operation is shown in Figure 9-21. All the other integer math instructions have the same input and output functions and format. A signal state of one (1) at the Enable (EN) input activates the math instruction. The instruction performs the math operation on the integer values at IN1 (input 1) and IN2 (input 2). If the result is outside the permissible range for an integer, the OV and OS bits of the status word are set to one and the Enable Output (ENO) bit of the instruction is set to zero. The result of the math operation is stored in the memory location programmed for the output (OUT) of the instruction.

Table 9-4. Integer Math Instructions

Instruction Symbol	Description
ADD_I	Add Integer
ADD_DI	Add Double Integer
SUB_I	Subtract Integer
SUB_DI	Subtract Double Integer
MUL_I	Multiply Integer
MUL_DI	Multiply Double Integer
DIV_I	Divide Integer
DIV_DI	Divide Double Integer

In the **Add Integer** instruction shown in Figure 9-21, a logic bit of one (1) at input I124.0 activates the ADD_I instruction. The result of the addition of word MW0 and word MW2 is entered into memory word MW10. If the result is outside the permitted range for an integer or the logic state of the Enable Input I124.0 is zero, the Enable Output (ENO) will be set to zero. However, if the result of the Add operation is valid and the Enable Input bit I124.0 is set to one, the output bit Q124.7 will be set to one.

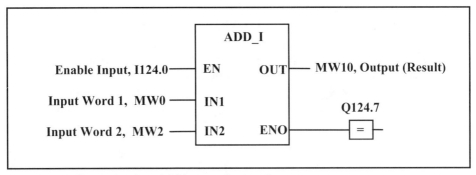

Figure 9-21. Example of an Add Integer instruction.

EXAMPLE 9-2

Problem: Write a Siemens Simatic S7 FBD program to add the integer data in word MW5 to the integer data in word MW7 and then multiply the result by 100, if the Enable Input bit I:124.0 is TRUE. Store the result in word MW100.

Solution: The function block diagram program to perform the math is shown in Figure 9-22.

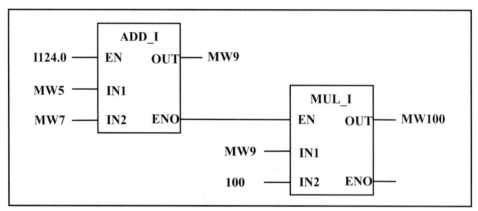

Figure 9-22. FBD program for math problem in Example 9-2.

Floating-Point Math Instructions

In this section, we will discuss the most common FBD instructions for performing mathematical operations on real numbers. The basic floating-point math instructions—add, subtract, multiply, and divide—are listed in Table 9-5. Consult the Simatic reference programming manual for a complete list of floating-point math instructions available in the S7 PLCs.

The floating-point math FBD instruction for an Add Real operation is shown in Figure 9-23. All the other floating-point math instructions have the same input and output functions and format. A signal state of one (1) at the Enable (EN) input activates the math instruction. The instruction performs the math operation on the values at IN1 (input 1) and IN2 (input 2). If either of the inputs or the result are not floating-point numbers, the OV and OS bits of the status word are set to one and the Enable Output (ENO) bit of the instruction is set to zero. The result of the math operation is stored in the memory location programmed for the output (OUT) of the instruction.

Table 9-5. Floating-Point Math Instructions

Instruction Symbol	Description
ADD_R	Add Real
SUB_R	Subtract Real
MUL_R	Multiply Real
DIV_I	Divide Integer

In the **Add Real** instruction shown in Figure 9-23, a logic bit of 1 at input I124.0 activates the ADD_I instruction box. The result of the addition of double-word MD0 and double-word MD4 is entered into memory double-word MD10. If either of the inputs or the result are not a floating-point number or if the logic state of the Enable Input I124.0 is zero, the Enable Output (ENO) will be set to zero. However, if the result of the Add operation is valid and the Enable Input bit I124.0 is set to one, the output bit Q124.0 will be set to one.

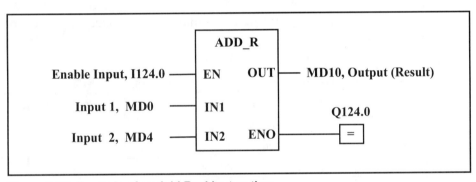

Figure 9-23. Example of an Add Real instruction.

EXAMPLE 9-3

Problem: Write a Siemens Simatic S7 FBD program to add the floating-point number in word MD0 to the floating point in word MD4 and then divide the result by 1.5 if the Enable Input bit I124.0 is TRUE. Store the result in word MD20.

Solution: The function block diagram program to perform the math is shown in Figure 9-24.

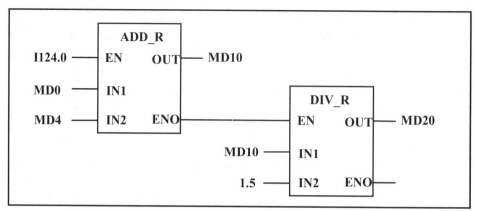

Figure 9-24. FBD program for math problem in Example 9-3.

Comparison Instructions

In this section we will discuss the three Siemens S7 comparison instructions: **Compare Integer**, **Compare Double Integer**, and **Compare Real**. The types of comparisons made in FBD programs are listed Table 9-6 along with the corresponding relational operator used in the compare instruction box.

Table 9-6. Types of FBD Comparisons

Types of FBD Comparisons	Relational Operator
Input 1 is equal to Input 2	==
Input 1 is not equal to Input 2	<>
Input 1 is greater than Input 2	>
Input 1 is less than Input 2	<
Input 1 is greater than or equal to Input 2	>=
Input 1 is less than or equal to Input 2	<=

Compare Integer

The **Compare Integer** instruction compares two values on the basis of sixteen-bit whole numbers. This instruction compares inputs IN1 and IN2 according to the type of comparison that is selected from the list of relational operators in Table 9-6. If the comparison is TRUE, the result of the operation is one (1). The comparison result cannot be negated, but the same effect as negation can be obtained by using the opposite compare function. An example of a **Compare Integer** is shown in Figure 9-25. If the value in word MW0 is equal to the value in word MW2 and the logic state of input I124.0 is one, then the output bit Q124.0 is set to one.

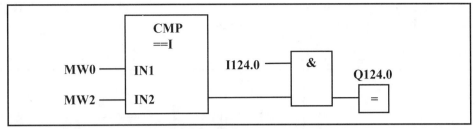

Figure 9-25. Example of an FBD Compare Integer instruction.

Compare Double Integer

The **Compare Double Integer** instruction compares two values on the basis of thirty-two-bit whole numbers. This instruction compares inputs IN1 and IN2 according to the type of comparison that is selected from the list of relational operators in Table 9-6.

If the comparison is TRUE, the result of the operation is one (1). The comparison result cannot be negated, but the same effect as negation can be obtained by using the opposite compare function. An example of a **Compare Double Integer** is shown in Figure 9-26. If the value in word MW0 is not equal to the value in word MW4 and the logic state of input I124.0 is one (1), then output bit Q124.0 is set to one.

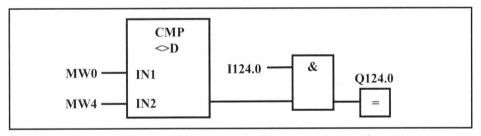

Figure 9-26. Example of an FBD Compare Double Integer instruction.

Compare Real

The **Compare Real** instruction compares two values on the basis of floating-point numbers. This instruction compares inputs IN1 and IN2 according to the type of comparison that is selected from the list of relational operators in Table 9-6.

If the comparison is TRUE, the result of the operation is one (1). The comparison result cannot be negated, but the same effect as negation can be obtained by using the opposite compare function. An example of a **Compare Real** is shown in Figure 9-27. If the value in word MW0 is greater than the value in word MW4 and the logic state of input I124.0 is 1, then output bit Q124.0 is set to one (1).

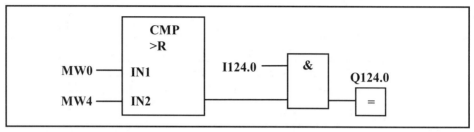

Figure 9-27. Example of an FBD Compare Real instruction.

EXERCISES

9.1   What are the four types of elements and boxes used in FBD programming?

9.2   Write an FBD program to start a pump that will fill a process tank with fluid until a high level is reached in the tank. Assume that a normally opened fill tank pushbutton is wired to a discrete input module at I124.4 and that a normally closed tank high-level switch is connected to input point I124.5 on a Siemens PLC model CPU314IFM. Also assume that the pump start relay is connected to output Q124.0 on the same Siemens PLC.

9.3   Write an FBD program to control an electric motor. Assume that a NO start pushbutton is wired to an input point I124.0 and that a NC stop pushbutton is wired to input I124.5 of a Siemens PLC model CPU314IFM. Also assume that output Q124.1 of the Siemens PLC will control a motor start relay and that there is an auxiliary motor start contact (NO) connected to the PLC input I124.1.

9.4   List the five timer instructions available in the Siemens S7-300 FBD programming instruction set.

9.5   Write an FBD program to control the temperature of the fluid in a process tank close to 400°C. Assume that the heater contactor is connected to PLC output Q124.1, a temperature-low switch set is connected to PLC input I124.0, and a temperature-high switch is connected to PLC input I124.1 on a Siemens PLC model CPU314IFM. The temperature-low switch is closed if the fluid temperature is below 395°C, and it opens at a temperature of 395°C or higher. The temperature-high switch is closed if the fluid temperature is below 405°C, and it opens at 405°C or higher temperatures.

9.6   Write an FBD program for a Simatic CPU314IFM PLC to activate a pump starter relay connected to output point Q124.0 if the inlet control valve FV-1 on a process tank is opened (input bit I124.0 =1) and the process tank is at a high level (input bit I124.2 = 1).

9.7   Write an FBD program for a Simatic CPU314IFM PLC that opens a fill valve on a process tank to allow an ingredient to be added to the tank for ten seconds each time the tank-level-high switch LSH-100 is activated (logic =1). Assume that switch LSH-100 is connected to input I124.0 and that the fill valve is wired to PLC output point Q124.0.

9.8   Write an FBD program for a Simatic S7-300 PLC that turns off a conveyor belt on a production line after fifty parts have been produced. Assume the following for the control program: (1) Output bit Q124.5 = 0 turns off the conveyor belt; (2) input I124.1 is set to one (1) each time a production part is rejected; (3) input I124.2 is set to one each time a new part is produced; (4) a normally open (NO) pushbutton connected to input I124.6 is used to set the production count to fifty, and (5) a NO pushbutton connected to input I124.7 is used to reset the counter to zero and to stop the conveyor belt.

9.9   Write an FBD program for a Simatic S7 that subtracts the integer data in word MW20 from the integer data in word MW18 and stores the result in word MW22 if the input bit I:124.0 is TRUE. Then divide the result by two, and store the final result in word MW24.

9.10   Write an FBD program for a Simatic S7 that subtracts a thirty-two-bit floating-point number in word MD70 from a thirty-two-bit floating-point number in word MD74 and stores the result in word MD78 if input I124.0 is set to one (1). Then divide the result by 4.5, and store the final result in word MD82.

BIBLIOGRAPHY

1.   Gilbert, R. A., and J. A. Llewellyn. *Programmable Controllers: Practices and Concepts* (Industrial Training Corp., 1985).

2.   Siemens AG. *Simatic S7 Function Block Diagram (FBD) for S7-300 and S7-400 Programming Reference Manual* (Siemens AG, October 1998).

10

Data Communication Systems

Introduction

The purpose of communications is to transfer information from one point or system to another. In process control, this information is called *process data* or simply, *data*.

To properly apply programmable controllers to process control and data collection it is essential to have an understanding of data communications. This chapter will provide a basic understanding of data communications terminology and concepts and their application to programmable controller systems.

Basic Communications

Data is transmitted through two types of signals: baseband and broadband. In a *baseband system*, a data transmission consists of a range of signals that are sent on the transmission medium without being translated in frequency. A telephone call is an example of a baseband transmission. A human voice signal in the 300-to-3,000-Hz range is transmitted over the phone line in that same frequency range. In a baseband system, there is only one set of signals on the medium at a time.

A *broadband transmission* consists of multiple sets of signals. Each set of signals is converted into a frequency range that will not interfere with other signals on the medium. Cable television is an example of broadband transmission.

As Figure 10-1 shows, three basic components are required in any data communications system: the *transmitter* to generate the information, the *medium* to carry the data, and the *receiver* to detect the data.

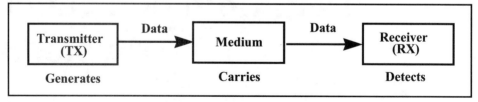

Figure 10-1. Basic communications system components.

The medium can be divided into more than one channel. A channel is defined as a path through the medium that can carry information in only one direction at a time. Figure 10-2 shows an example of a multichannel medium with n channels. Probably the most common example of multichannel communications systems are cable television systems.

Physical transmission media fall into four general categories: multiconductor cable, twisted-pair cable, coaxial cable, and fiber-optic cable. *Multiconductors* consist of two or more insulated electrical conductors inside a plastic-covered cable. A typical example is the parallel printer cable used between a personal computer and a printer. *Twisted-pair cables* consist of two electrical conductors, each covered with insulation. The two wires are twisted together to ensure that they are both equally exposed to electrical interference signals in the environment. Since the wires are carrying current in opposite directions, the electrical interference will tend to cancel out in the cable. Twisted-pair is the most common cable used in programmable controller systems. It is the least expensive transmission medium and provides adequate electromagnetic interference (EMI) immunity.

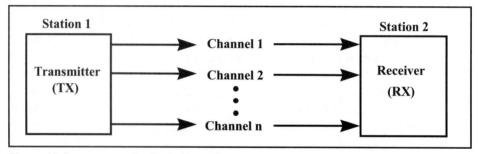

Figure 10-2. n-communication channels.

A *coaxial cable* consists of an electrical conductor surrounded first by insulating material and then by a tube-shaped metal braid conductor. In most cases, the entire cable is covered by an insulator. The round conductor in the center of the cable and the circular outer tube conductor are coaxial in that they share the same central axis.

Coaxial cables are used in both baseband and broadband communications systems. A variety of cable types are used in process control applications, ranging from flexible types similar to the coaxial cables used on home TV sets to heavy, rigid cables that require special installation procedures. Coaxial cables are generally used in process automation applications involving long communications runs of over 1,000 feet and where improved EMI immunity is required. Coaxial cable finds very extensive use in plantwide communications networks.

The *fiber-optic cable* consists of fine glass or plastic fibers. At one end, electrical pulses are converted into light by a photodiode and sent down the fiber-optic cable. At the other end of the cable, a light detector converts the light pulses back into electrical signals. The light signals can travel only in one direction, so two-way transmission requires two separate fiber cables. A fiber-optic cable is normally the same size as a twisted-pair cable, and it is immune to electrical noise. The cost of fiber-optic cables is about the same as coaxial cables. However, the use of fiber optics in programmable controller applications has been limited by the high cost of connectors and the lack of industrial standards.

Communications can also be described in terms of the number of channels used for information flow. The three common methods of data transmission are unidirectional, half-duplex, and duplex. We will discuss each in the following sections.

Unidirectional Communications

In *unidirectional communications*, a single channel is used and communication is only one-way (i.e., from transmitter to receiver), so the receiver can never respond. In unidirectional communications, it is not possible to send errors or control signals from the receiver station because the transmitter (TX) and the receiver (RX) are each dedicated to performing one function. A typical example of unidirectional communications in process control would be a weigh scale in the field sending data to a programmable controller in a central control room.

Half-Duplex Communications

Two-way communication allows the receiver to verify that the data was received. One kind of two-way communication is called *half-duplex*. In half-duplex, a single channel is used, and communication is two-way, but it can occur only in one direction at a time (see Figure 10-3). In this configuration, the receiver and transmitter alternate functions so communication occurs in one direction at a time on the single channel. A circuit in the station communications module acts as an OR gate to place the unit in transmit or receive mode.

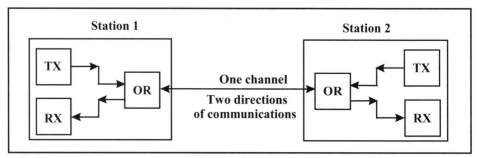

Figure 10-3. Half-duplex communications.

Full-Duplex Communications

Two-way communications in which data can flow in both directions at the same time is called *full-duplex communications*. In this case, two physical paths are required so information can flow from both directions simultaneously, as shown in Figure 10-4.

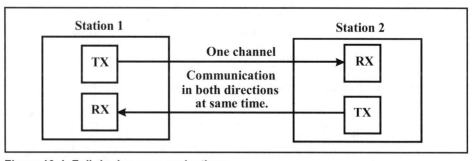

Figure 10-4. Full-duplex communications.

Transmission Methods

Transmission involves moving data between a receiver and a transmitter. The two methods of transmission are *parallel* and *serial*.

Parallel Transmission

In parallel transmission, all bits are transmitted at the same time, and each bit of information requires a unique channel. For example, if we transmit an eight-bit ASCII character, eight channels are required. Figure 10-5 demonstrates the transmission of the ASCII character using odd parity. The term *parallel* refers to the position of the bits of the character and the fact that the characters are transferred as a group of bits in parallel. The use of a group of channels results in a high data-transfer rate.

Parallel data transfer does carry with it the problem of bit synchronization decay. Synchronization is the process of making two or more activities happen at the same rate and time. If we send a string of characters on a long parallel cable, the difference in impedance in the cable wires can cause bit synchronization decay, as shown in Figure 10-6.

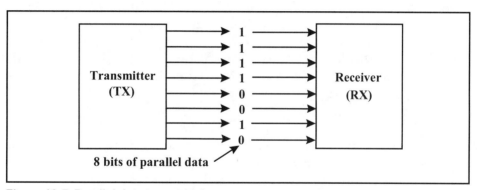

Figure 10-5. Parallel data transmission.

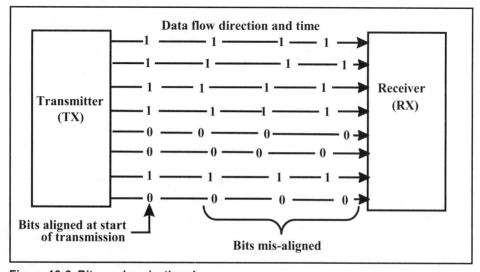

Figure 10-6. Bit synchronization decay.

To avoid the transfer errors caused by bit synchronization decay, the distance between the receiver and the transmitter must be kept short. For this reason, parallel transmission is used only in closely connected data systems where high speed is required. For example, data transfer on data and address buses in a microprocessor is parallel transmission. However, in process control systems, the data transfer between systems such as personal computers and programmable controllers is normally serial. For this reason, we will limit our discussion in the remaining parts of this chapter to serial data transmission.

Serial Transmission

In this section, we will discuss serial interfacing between microcomputer-based devices. Serial interfacing is probably the least complex electrical interface to implement because it requires only one signal wire to carry all data flow in one direction and only two wires for bidirectional flow. Serial interfaces do, however, require additional logic circuitry for converting between parallel and serial data because most computers process data in parallel. This extra circuitry is not required in the parallel transmission of information. Because serial links are low in cost and easy to install, they have been standardized into a few widely used protocols (transmission control standards) where several low-cost integrated circuit (IC) interface chips are available.

In serial transmission, the bits of the encoded character are transmitted one after another in a single channel (see Figure 10-7). The transmission takes the form of a bit stream that the receiver must assemble into characters (normally eight bits) using specially designed ICs.

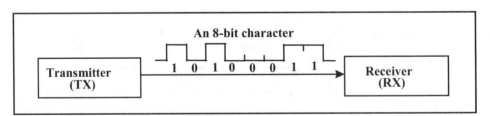

Figure 10-7. Serial eight-bit character transmission.

The main advantages of serial communications are lower cost and the elimination of byte synchronization problems. Controlling the speed of transmission is critical in serial data transfer. A common measurement unit used to describe serial data speed is *bits per sec (bps)*.

The process of bit synchronization is required for serial data transfer. A timing circuit at the transmitter places transmitted bits on the channel at fixed intervals determined by the selected communication rates. Some communications rates typically used are 2.4K, 4.8K, 9.6K, 14.4K, and 28.8K bps.

The transmitter must send bits at the same rate that the receiver is set to accept them. For example, if a personal computer is transmitting serial data at 9,600 bps to a programmable controller system, the communications module on the programmable controller must be set at a communications rate of 9,600 bps.

Character synchronization is used to determine which eight consecutive bits in a data stream represent a character. The receiver must recognize the first data bit and count bits until the character or byte is complete. Various forms of character synchronization are used in data communications. Sometimes each byte is framed with special bits to indicate where to start and stop.

Signal Multiplexing

A single line can be used to carry more than one signal by using signal multiplexing. To understand multiplexing, we must first define the term *bandwidth*. Bandwidth describes the signal-carrying capability of a communications channel or line. Bandwidth is defined as the difference in cycles per second between the lowest and highest frequencies that a channel can handle without a certain amount of signal loss.

Multiplexing allows one line to serve more than one receiver by creating slots or divisions in the line. The equipment used to obtain multiplexing is called a multiplexer or MUX (see Figure 10-8).

The commonly used signal multiplexing methods are frequency division, time division, and statistical multiplexing.

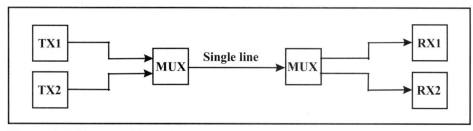

Figure 10-8. Signal multiplexing.

Frequency Division Multiplexing

In frequency division multiplexing, each channel is assigned its own individual frequency range. Frequency division is normally used to combine a group of low-speed sources onto a single voice line. A typical application of frequency division communications is shown in Figure 10-9.

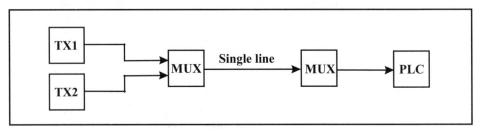

Figure 10-9. Frequency division multiplexing.

Time Division Multiplexing

Time division multiplexing uses time periods to assign channel space and allot the available bandwidth. In Figure 10-10, two time periods (Period 1 and Period 2) are used to assign each channel its own individual time slot in the data stream. In this type of multiplexing, no other channel can use the time slot, so there is some wasted bandwidth when a station is not transmitting data in its time slot. However, one transmission line can support many data streams. Time division multiplexing is normally used where there is a need to combine a number of relatively low-speed data transmissions onto a single high-speed line.

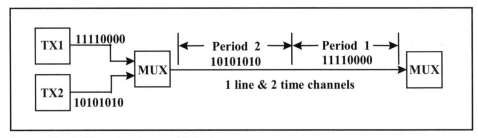

Figure 10-10. Time division multiplexing.

Statistical Multiplexing

Statistical multiplexing is an enhancement of time division multiplexing designed to reduce the wasted bandwidth when data is being transmitted by a station. This enhancement process is sometimes called *data concentration*. Statistical multiplexing is generally used where a large number of data entry terminals require only a brief or occasional data transfer. If the use of the terminals becomes intensive, the operators will experience long time delays, so this method must be applied with caution.

Error Control and Checking

Error control and checking is the process used to detect any discrepancy between transmitted data and received data in a communications system. Errors detected at the receiver are either corrected or retransmitted. The common error-control methods used in programmable controller systems are (1) echo check, (2) vertical redundancy check (VRC), (3) longitudinal redundancy check (LRC), and (4) cyclical redundancy check (CRC).

Echo Check

Echo check is used in two-way communications systems (Figure 10-11) to check the accuracy of transmitted data.

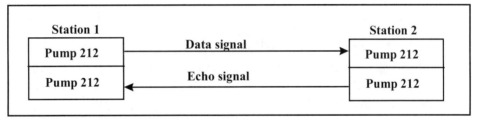

Figure 10-11. Echo check.

The receiver (RX) in Station 2 sends every character it has received back to the originating terminal (Station 1). Station 1 compares the echoed data with the information it sent. The comparison is performed by electronic circuitry in the transmitter/receiver. If a transmission error is detected, the information is retransmitted a fixed number of times to obtain an error-free transmission.

Vertical Redundancy Check (VRC)

In the *vertical redundancy check* (VRC) process, parity bit checking is used to detect the change of a single bit. In this method, a single bit is added to a character string to create either an odd or an even number of 1 bits. Parity bits are often called "redundant" because they can be removed without loss of data. If even parity is used, a 1 bit is added to a character string to make the total of 1 bits even. For example, the ASCII letter *A* is given by 0000001 (see Table 10-1). If even parity is used, a 1 bit is added to the binary string before transmission (i.e., *A* = 10000001). If odd parity is employed, a 0 bit is added to the binary string to make the eight-bit binary string $A = 00000001$.

In even-parity VRC transmission, the receiver would detect an error when it receives a character that contains an odd number of one(1) bits. In this method, after a parity error is detected, the receiver would request retransmission of the data.

Table 10-1. ASCII Code

Bits	7 6 5	0 0 0	0 0 1	0 1 0	0 1 1	1 0 0	1 0 1	1 1 0	1 1 1
4321	HEX	0	1	2	3	4	5	6	7
0000	0	NUL	DLE	SP	0	@	P	'	p
0001	1	SOH	DC1	!	1	A	Q	a	q
0010	2	STX	DC2	"	2	B	R	b	r
0011	3	ETX	DC3	#	3	C	S	c	s
0100	4	EOT	DC4	$	4	D	T	d	t
0101	5	ENQ	NAK	%	5	E	U	e	u
0110	6	ACK	SYN	&	6	F	V	f	v
0111	7	BEL	ETB	'	7	G	W	g	w
1000	8	BS	CAN	(	8	H	X	h	x
1001	9	HT	EM	)	9	I	Y	i	y
1010	A	LF	SUB	*	:	J	Z	j	z
1011	B	VT	ESC	+	;	K	[	k	{
1100	C	FF	FS	'	<	L	\	l	|
1101	D	CR	GS	-	=	M	]	m	}
1110	E	SO	RS	.	>	N	^	n	~
1111	F	SI	US	/	?	O	-	o	DEL

Example 10-1 will help to explain the concept of parity.

EXAMPLE 10-1

Problem: Generate the ASCII code with even parity for the word *stop*.

Solution: Using the ASCII code in Table 10-1, we first obtain the binary code for each character as follows:

Bits	7	6	5	4	3	2	1	Character
	1	0	1	0	0	1	0	S
	1	0	1	0	1	0	0	T
	1	0	0	1	1	1	1	O
	1	0	1	0	0	0	0	P

Next, we count the number of one bits for each character. If the number is even, we set the parity (P) bit to zero to obtain an even number of one bits in the binary string. On the other hand, if the number of one bits is odd, we set the parity bit to one to obtain an even number of 1 bits in each string, as follows:

Bits	P	7	6	5	4	3	2	1	Character
	1	1	0	1	0	0	1	0	S
	1	1	0	1	0	1	0	0	T
	1	1	0	0	1	1	1	1	O
	0	1	0	1	0	0	0	0	P

Longitudinal Redundancy Check (LRC)

Longitudinal redundancy check (LRC), or block check character (BCC), is a procedure that checks an entire horizontal line within a block of data for odd or even parity. This process is used in combination with VRC. While VRC can detect a single error, the only way to obtain corrected data is by retransmission. When LRC is used with VRC, any single-bit error in an entire data block is not only detected but can also be corrected at the transmitter without a retransmission. This increases the overall data transmission speed for a system.

To illustrate the longitudinal redundancy check process, let us assume we are transmitting the data word *run* in ASCII code using odd parity, as shown in Figure 10-12.

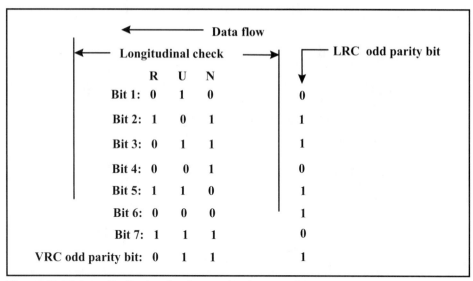

Figure 10-12. Longitudinal redundancy check example.

In Figure 10-12, the transmitter sends an LRC odd-parity character. This character is generated by adding a parity bit at the end of the entire transmitted block so an odd number of one bits is created for each longitudinal row of bits.

At the receiver, the LRC is calculated for the data bytes in the block, and it is compared with the transmitted LRC character. If they are not equal, the vertical parity bit reveals which byte is in error, and the LRC reveals which of the eight bits is in error. Logic circuitry or software in the receiver is used to change the error bit to its opposite state, correcting the error. Only if there is more than one error in the block transfer must retransmission be requested by the receiver.

EXAMPLE 10-2

Problem: Calculate the VRC/LRC odd-parity characters for the word *pump*.

Solution: Using the ASCII code in Table 10-1, first obtain the binary code for each character as shown in the table. Next, count the number of one bits for each character. If the number is odd, set the vertical parity bit (bit 8) to zero to maintain an odd number of one bits in the binary string. On the other hand, if the number of one bits is even, set the parity bit to one to obtain an odd number of one bits in each vertical string, as shown in Table 10-2.

> **EXAMPLE 10-2 continued**
>
> Finally, count the number of one bits in each row to generate the longitudinal redundancy check (LRC) odd-parity bit. If the number of ones is odd, set the LRC parity bit to zero to maintain an odd number of one bits in the longitudinal row. On the other hand, if the number of one bits is even, set the LRC parity bit to one to obtain an odd number of one bits in each row, as shown in Table 10-2.

Table 10-2. VRC/LRC Odd-parity Example

Word	P	U	M	P	LRC Parity
Bit 1	0	1	1	0	1
Bit 2	0	0	0	0	1
Bit 3	0	1	1	0	1
Bit 4	0	0	1	0	0
Bit 5	1	1	0	1	0
Bit 6	0	0	0	0	1
Bit 7	1	1	1	1	1
Vertical Parity Bit	1	1	1	1	0

Cyclical Redundancy Check (CRC)

Cyclical redundancy check (CRC) is a method for checking errors on an entire data block. In this method, a check character is transmitted at the end of each block. The transmitter calculates the check character from the data transmitted. The receiver compares the transmitted CRC to its own calculated CRC, based on the received block of data. If they are not equal, the receiver requests retransmission of the previous message block. If they are equal, the transmitted data is assumed to be error free.

Cyclical redundancy check is a very accurate method for detecting data transmission errors. It is the most common method used in programmable controller-based systems.

Communications Protocols

Communications protocols are a set of rules that specify the format and control of transmission between two communications devices. The two main functions of protocols are *handshaking* and *line discipline*. Handshaking determines whether a circuit is available and makes sure the circuit is ready to transfer data. Line discipline performs the following functions: (1) receiving and transmitting information, (2) error control procedures, (3) sequencing the message blocks, and (4) error checking.

To illustrate the concept of communications protocols, Figure 10-13 shows a typical line discipline sequence between a transmitter and receiver. The two basic transmission methods used to obtain line discipline are *asynchronous* and *synchronous*. In asynchronous, successive data appear in the data channel at arbitrary times with no specific clock control governing the time delays between information. In synchronous, each successive datum in a data stream is controlled by a master data clock and appears at a specific interval of time.

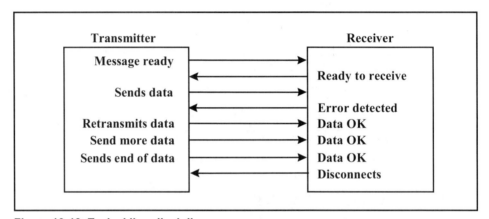

Figure 10-13. Typical line-discipline sequence.

Most synchronous and asynchronous systems deliver serial data in eight-bit characters. The asynchronous systems treat each character as an individual message, and the characters appear in the data stream at arbitrary, relative times. However, within each character, the eight bits are transmitted at a fixed predetermined clock rate such as 2,400, 9,600, 14.4K, 28.8K, and 33.6K bps. Hence, an asynchronous communications system is actually synchronous within a character and asynchronous between characters.

A typical bit stream for an asynchronous transmission is shown in Figure 10-14. Line discipline is achieved by framing every character transmitted with start/stop bits. The start bit is a zero or "space" and the stop bits are a one or "mark." Asynchronous transmission operates at random speeds and is almost always used with half-duplex protocols.

Synchronous transmission sends an entire message at one time and uses special character strings to achieve line discipline. It includes a predefined start and end sequence and normally operates at speeds of 2,400 bps and

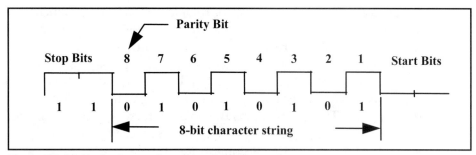

Figure 10-14. Typical asynchronous character string.

higher. A typical sequence begins with two synchronous (syn) characters, as shown in Figure 10-15.

In Figure 10-15, the start-of-message (SOM) character indicates the beginning of the data transfer, and the control character gives the receiver control information on the data transfer. The actual data can vary in length and is followed by error checking, which is normally CRC. The end-of-message (EOM) is used to signal the end of a data transmission.

The two transmission modes used for line discipline are half-duplex and full-duplex protocols. Half-duplex protocol allows communications in only one direction. In a two-wire half-duplex circuit, the stations must be able to switch automatically from transmit to receive. In a four-wire full-duplex circuit, each station has a dedicated receiver and transmitter circuit, but data cannot be transmitted and received at the same time. Full-duplex protocol permits simultaneous transmissions in both directions. It falls back to half-duplex operation if the communications link cannot support full-duplex.

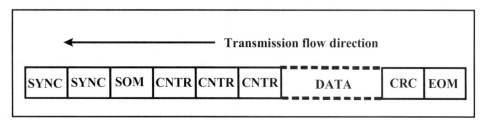

Figure 10-15. Typical synchronous transmission.

Serial Synchronous Transmission

Synchronous communications depend on the data and protocol control information being assembled into a structured and predefined package or format. The format tells the computer or other equipment what to do with the message and how to do it.

A message will contain one or more fields of data. Each synchronous transmission in a message block format will contain three major parts: a header, text, and the trace or trail block, as shown in Figure 10-16. The *header* starts the message package and makes it possible for the transmitter and receiver to synchronize. It normally contains characters that identify the transmitting or receiving stations and provide communications routing data. The *text* contains the data characters to be transferred and consists of a singleblock or multiblock message. The *trace* or *tail* characters signal the end of a transmission block.

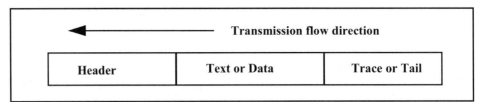

Figure 10-16. Three elements of a typical synchronous transmission.

Serial Synchronous Protocols

Different protocols use different control and data formats. A communications protocol can be defined as a fixed set of rules governing the format and control of inputs and outputs between two data transfer devices. In a standard data transfer system, a protocol governs the following: line control, framing, error control, and sequence control.

Line control is used to list the station that will transmit and the station that will receive in half-duplex communications. It is also required in a multipoint circuit (a circuit connecting three or more points on a common line). The *framing* function normally determines which eight-bit groups are the characters and which groups of characters are the messages. *Error control* detects transmission errors using various redundancy checks and normally corrects faulty messages. *Sequence control* function numbers the messages to eliminate duplication and avoid data loss. It also identifies any retransmitted messages.

There are basically three serial data protocols in wide use today in process control:

1.   BISYNC (binary synchronous communications), an older protocol used in IBM equipment

2.   DDCMP (digital data communication message protocol), a protocol used primarily by Digital Equipment Corporation (DEC)™ equipment

3.   HDLC (high-level data-link control), probably the most widely used protocol

Because BISYNC and HDLC protocols are the most widely used in programmable controller systems, the discussion will be limited to these two techniques. HDLC has evolved from two earlier standards, the ADCCP (advanced data communications control procedure) and the SDLC (synchronous data link control). Because of the close similarity of the three, the discussion will focus only on the HDLC protocol. But first, the BISYNC protocol needs to be discussed.

BISYNC Protocol

BISYNC stands for *binary synch*ronous communications; it is a half-duplex character-oriented protocol. The protocol has a very rigid format that uses special characters (ASCII or EBCDIC-Extended Binary Coded Decimal Interchange Code) to delineate the various fields of a message and to control the required protocol functions.

A typical BISYNC message, shown in Figure 10-17, consists of the following discrete parts: (1) two or more synchronizing characters (SYNC), (2) start-of-header (SOH), (3) header, (4) start-of-text (STX), (5) text, (6) an end-of-text (ETX), and (7) a trace or tail block (CRC).

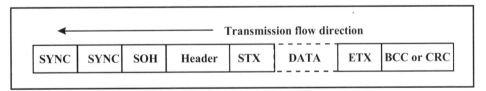

Figure 10-17. A BISYNC transmission string.

The *synchronizing (SYNC) characters* are used to establish the correct timing between the transmitter and receiver. The number of SYNC characters varies with the different communications applications and networks. The *start-of-header* (SOH) is a format control character that is transmitted just before the header character to identify the individual message control characters. The *header* is an optional character that normally contains routing or message priority information. The *start-of-text* (STX) is a special format control character that is transmitted before the first data characters; it indicates that the characters to follow are information. The *text*, of course, is the data that is being transmitted. The end-of-transmission block (ETB) is a format control character indicating the end of text and the beginning of the trace or tail (CRC). It is normally used to indicate the end of an intermediate text block.

The *end-of-text* (ETX) is also a special format control character that indicates the end of a text block and the start of the trace or tail (CRC). The *trace or tail block* (CRC) detects and corrects errors in transmission. It depends on the information code being used, such as ASCII or EBCDIC, and it has a block check character or a combination of checks.

If the trace or tail block is in ASCII, a VRC/LRC message check is performed. In EBCDIC transmission, normally no VRC/LRC check is performed, and the CRC is calculated on the entire message.

The BISYNC is a rather simple protocol, but several problems complicate it. For example, the format of the protocol places special meaning on the ETX character. If the data block (or control information) contains this character among its data, the characters can be misinterpreted. For example, if the datum happens to be an eight-bit pattern identical to the ASCII pattern for ETX, this datum character could deceive the receiver into taking an end-of-block action when there are actually more characters to follow in the message block. To correct this problem, the protocol needs to distinguish specific patterns as data characters. The ability to treat control characters either as control information or as data is called *data transparency.*

BISYNC uses the control character DLE (data link escape) to obtain the required transparency. If a control symbol is to be treated as data, it is preceded by the DLE character. In other words, by receiving a DLE the receiver is warned to accept the next character as data and not to take any control action. Using the DLE character is somewhat more complicated than this because of other special considerations. For example, to maintain transmission synchronization in the absence of data in the transmitter queue, the protocol provides for the automatic insertion of sync characters, which are ignored by the receiver. But it is possible that a sync character might be inserted between a DLE and a control character that follows it. This condition forces the receiver to interpret the sync character as if it were data rather than accept the next character as data. In this situation, the transmitter cannot simply send a sync character after a DLE. The transmitter must queue the DLE and then send sync characters until both the DLE and its corresponding data character are ready.

If message buffering is not available, the transmitter has to send a DLE and then stay idle on the line. In this situation, it should transmit in the idle state the characters *DLE* and *SYNC*, which the receiver can ignore as pairs of characters. Also note that since the DLE is a control character with special significance, the DLE character in the data transmission must be treated the same way that the ETX is treated, and it must be preceded by a

DLE character when it is transmitted over a communications link. Example 10-3 will help to illustrate the BISYNC protocol.

EXAMPLE 10-3

Problem: Assemble the bit streams that are required to send the message M1 ON using eight-bit odd-parity ASCII code and BISYNC protocol. Assume that the header is given by 00000001 for transmitting station 1, and omit the check character (BCC or CRC) at the end of the transmission.

Solution: Use the ASCII code in Table 10-1 to find the bits required as follows: SYNC = 00010110, SOH = 00000001, STX = 00000010, M = 11001101, 1 = 00110001, space = 00100000, 0 = 01001111, N = 11001110, ETX = 10000011.

The HDLC Protocol

The main feature of HDLC protocol is that it opens and closes each message block or frame with start-frame and stop-frame characters (flags). A typical HDLC message is shown in Figure 10-18. It consists of six discrete parts: (1) open flag, (2) address byte, (3) control byte, (4) data field, (5) a check field, and (6) a close flag. The *open flag* always consists of the same eight bits (01111110); it is used to indicate the start of a transmission frame. This eight-bit sequence is never repeated throughout the entire message until the close flag.

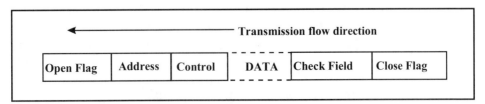

Figure 10-18. An HDLC transmission string.

The *address byte* is the address of the transmitter on a command message or the receiver on a response message. The address byte consists of eight bits, thus allowing for 256 addresses. However, 16 addresses are the maximum normally used since some bits are used for other functions.

The *control byte* has eight bits and contains command or control response information. The *data field*, on the other hand, can contain any number of bytes and is the user's total data transmission. The data field normally uses EBCDIC, ASCII, BCD, or straight binary code. The *check field* follows the data and precedes the close flag. It contains a cyclical redundancy check (CRC) character that detects and in some cases corrects errors during transmission.

To review, CRC functions as follows: the transmitter sends its computation to the receiver, which compares the transmitted computation with its own calculation. If they are equal, the data is assumed to be error-free and, if unequal, the receiver may not accept the transmission and normally requests retransmission.

The *close flag* is the final transmission byte and has the same bit configuration as the open flag. It terminates the transmission frame and begins the next if it is available.

The synchronous nature of this protocol forces the transmitter to have data ready in a buffer at the beginning of a transmission block. If it is not ready and the system fails to produce the data in time for transmission, the transmitter will run out of information to transmit. The HDLC protocol does not have an idle character within a block, so the system must abort an entire transmission block when the transmitter runs out of data to send. The abort code is normally a sequence of eight 1s.

Local Area Networks

A local area network (LAN) is a user-owned and -operated data transmission system that operates within a building or set of buildings. LANs allow a great number and variety of machines and processes to exchange large amounts of information at high speed over a limited distance. LANs connect communicating devices, such as computers, programmable controllers, process controllers, terminals, printers, and mass storage units, within a single process or manufacturing building or plant.

LANs allow neighboring computers to share data resources, hardware resources, and software. For example, a typical manufacturing facility might tie together the process control system computers and a central computer used by purchasing to speed up the ordering of raw materials for the plant.

A typical LAN for an industrial facility is shown in Figure 10-19. It consists of three levels with three different types of networks. The highest level (Level I) is the information network used by groups like accounting and purchasing at the plant. This network has the highest-speed communications (10 Mbps and higher) network because it must handle large amounts of data and information. The middle level (Level II) is used to perform the automation and control of the industrial plant. This network is generally slower (normally 100 Kbps to 5 Mbps) than the

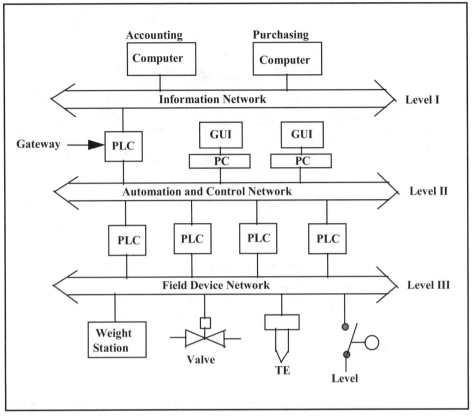

Figure 10-19. Typical industrial plant LANs.

highest level network because the data rates required for process and machine control are lower. The low-level bus (Level III) is used to connect programmable controllers directly to the field devices, such as weigh, flow, pressure, level transmitters, and switches.

LAN Topologies

There are three LAN topologies in common use: ring, bus, and star. The *ring* network topology shown in Figure 10-20 can operate using a unidirectional transmission medium. However, most ring topology networks use bidirectional transmission to allow messages to flow with the most efficiency. Each node decides whether to accept or pass on a message. This scheme is relatively easy to implement.

The *bus* network topology shown in Figure 10-21 requires a broadcast medium in which signals flow to all stations at all times. All the stations receive transmissions, even if they act only on some. The advantage of this scheme is that the stations connected to the bus perform no message routing because the bus is a broadcast type medium. This is the topology most commonly found in process control.

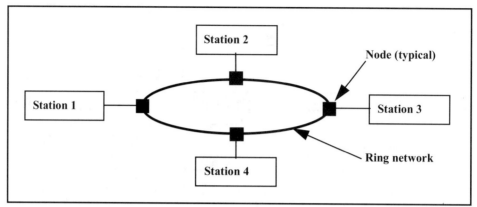

Figure 10-20. Ring-type LAN.

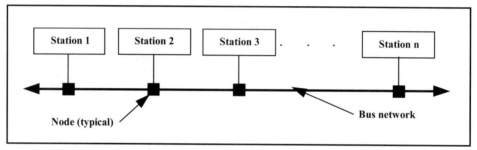

Figure 10-21. Bus-type LAN.

The *star* network shown in Figure 10-22 normally allows only one station to be in communication with the central station at one time. The central station may be allowed to transmit to several nodes at the same time. Routing messages is very easy because the central station has a unique hardware path to each node. System security is high since access to the network is controlled by the central station. Another advantage is that priority status can be assigned to selected nodes in the network.

LAN Protocols

Earlier, we defined a communications protocol as the set of rules that govern data communications. One of the most important functions of a LAN protocol is governing access to the communications network. Failure to control access will result in chaos any time the traffic on the network rises above a certain minimum level.

The standard LAN protocols are *polling*, *token passing*, and *carrier sense multiple access/collision detect* (CSMA/CD). In polling, a master station selects each of the other stations in turn and gives the station permission to communicate for a fixed period of time. The main disadvantage of polling

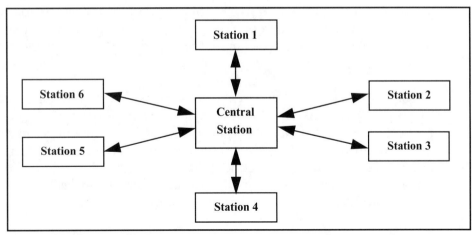

Figure 10-22. Star-type LAN.

is that each station must remain idle until it is selected by the master. This type of protocol is best suited for bus- or star-type LAN topologies, but it is not normally used because stations must remain idle a large percentage of the time.

The token-passing protocol operates by passing a symbolic electronic token from one station to another in the network. Each station may hold the token for a predetermined length of time before passing it. The station that has the token controls the right to transmit to the network.

The carrier sense multiple access/collision detect (CSMA/CD) protocol allows stations to try to communicate whenever they need access. When a station has a message to send, it first listens to determine if anyone else is transmitting ("carrier sense"). When the station detects an idle channel, it transmits data. If two stations detect an idle channel and transmit simultaneously, a collision will occur. In this case, the "collision detect" part of the protocol informs both stations that the communication has failed; then both stations wait a random amount of time before attempting to retransmit.

Standard Network Architecture

Data communications between computer systems and networks is possible only if they adhere to some common set of rules governing both hardware and software. A prerequisite of computer system design is a standard approach to network design or architecture that defines the relations between network services and functions.

The International Organization for Standardization (ISO) recognized the need for standards to govern the information exchange between and within networks and across geographical boundaries. Their standard, which has gained wide acceptance, is a seven-layer model for network architecture known as the ISO model for Open System Interconnection (OSI).

The ISO/OSI's layered approach to network design is based on the design of operating systems. Due to the complexity of most computer operating systems, they are generally designed in sections, each of which contributes a certain function to the operating system. This method of design makes it easier for each section to be refined and redesigned independently to meet its functional purpose. Finally, all of the layers are integrated to provide a totally functioning operating system.

The same procedure can be used to design a network system. The ISO/OSI model specifies a hierarchy of independent layers that contain modules for performing defined functions. The ISO/OSI model has seven distinct layers at both the receiver and the transmitter through which communications must pass, as shown in Figure 10-23.

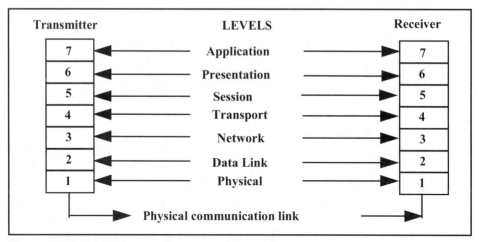

Figure 10-23. ISO/OSI seven-layer communications standard.

The function of each layer of the ISO/OSI model is described as follows:

1.   The *physical* layer defines the electrical and mechanical requirements of interfacing to a physical medium in order to transmit information. When used, this layer must include the software driver for each communications device as well as the hardware, such as interface equipment, the connectors, modems, and the communication cables.

2.   The *data link* is used to establish an error-free communications link between network stations over the physical channel. It formats messages for transmission, checks the integrity of received data, controls access to and use of the station, and ensures the proper sequence of the transmitted information.

3.   The *network* control layer is used to address messages, set up the path between stations, route messages across intervening stations to their destinations, and control the flow of data between stations.

4.   The *transport* layer furnishes end-to-end control of a communication once the path has been established. This allows the system to exchange data reliably and sequentially. This layer is normally beyond the user's control.

5.   The *session* layer organizes the dialog of the communication and manages data exchange. This layer is also normally beyond the user's control, as are layers 6 and 7.

6.   The *presentation* layer handles tasks related to data representation and code conversion.

7.   The *application* layer handles tasks related to data transfer speed and integrity.

Outside the control of the ISO/OSI are the communications applications processes. Usually, one application process is at the transmitter and another is at the receiver. Typical examples of application processes are computer operations that require a user to be at a terminal or require a piece of software to perform instructions.

Serial Hardware Standards

Serial hardware standards make possible the physical and electrical compatibility among devices that communicate. Ports and connectors must match in size, shape, and pin configuration. Transmitting and receiving circuits must agree on how signals are to be generated and interpreted. The so-called physical-layer communications requirements are handled by the serial interface standards, such as EIA232, RS-422, RS-485, and 20-mA current loop.

The Electronics Industries Association (EIA) is responsible for setting the physical and electrical standards in the United States. There are many different communications standards because there are so many different purposes and functions to be served.

The term *serial interface* or *interface* is often used to mean a specific physical and electrical standard that applies to a given piece of equipment. For example, if we say that a programmable controller has an "EIA232 serial interface," we mean that the interface is the circuitry that allows the PLC to communicate with another device. This interface circuit and its associated software is responsible for the ISO/OSI layer 1 physical and electrical characteristics of the interface device as well as the required layer 7 application functions. On the other hand, when some people refer to an EIA232 interface they simply mean the port or the connector.

EIA232 Standard

The EIA232 standard is one of the most common serial interfaces in use today. It defines a number of physical and electrical characteristics. Most EIA232 connectors have 25 connection pins and are shaped like a *D*. Each pin, or data line, is assigned a special purpose. In this standard, a logical 1 is a voltage greater than +3 volts dc; logical 0 is a voltage less than -3 volts dc.

Data transfer rates up to 25 kilobaud are possible with this standard, but cabling between devices is limited to 50 feet. EIA232 is a master/slave serial interface, that is, there is a primary station in control of a secondary station or stations. The term *unbalanced link* is also used to mean EIA232 communications.

RS-422 Standard

The RS-422 interface standard, unlike EIA232, creates balanced link communications. It uses two data lines, line A and line B, each of which transmits and receives. The voltage value of the data signal is determined by the difference in voltage between the two lines sensed by the interface electronic circuitry.

RS-422 is much faster than EIA232, and transmission rates up to 10 megabaud are possible. The interface can support up to 32 drops along the network. Wiring or cabling distances are much longer than with EIA232— up to 4,000 feet. This is because the voltage difference, not its specific value, determines whether a signal is one or zero.

RS-485 Standard

The RS-485 interface standard is based on impedance sensing rather than voltage sensing. Like the RS-422 interface, RS-485 can be run for distances up to 4,000, and it can support up to 32 device connections.

The wiring guidelines under RS-485 are very strict. For example, all connections are polarized, and cables must be kept a specific minimum distance from power cables and wires. Where power wires cross an RS-485 cable, the crossing must be made at 90° angles. The RS-485 standard has greater immunity to noise interference than either EIA232 or RS-422.

20-mA Current Loop Standard

The 20-mA current loop standard is very popular in industrial control system applications. The current in the serial communications line is kept at a constant 20 milliamps. Logic one and logic zero are determined by opening and closing the current loop between devices in the network. Current ON represents logic one, and current OFF represents logic zero. Since the current is kept constant at 20 mA for logic one, the resistance (i.e., length of the communications line) has no effect on the data signal. This means that long network distances are possible.

EXERCISES

10.1   What are the three basic components of a communications system?

10.2   Explain how unidirectional, half-duplex, and full duplex communications operate.

10.3   Compare and discuss serial and parallel data transmission. Describe the advantages of each method and some typical applications of them that might be encountered in computer systems.

10.4   What are the three most common methods of signal multiplexing encountered in communications systems?

10.5   What are the four most common data transmission error-checking methods used in PLC communications systems?

10.6   Calculate what VRC/LRC odd-parity characters are required for the message M1 ON.

10.7   Explain the difference between asynchronous and synchronous data transmission.

10.8   List three common serial data protocols used in communications systems.

10.9   Assemble the bit streams required to send the message RUN using eight-bit even-parity ASCII code and BISYNC protocol. Assume that the header is given by 00000010, which represents station 2, and omit the check character at the end of the transmission.

10.10   List the seven layers of the ISO/OSI communications standard.

BIBLIOGRAPHY

1. Allen-Bradley Co., Inc. *Reference Manual Data Highway/Data Highway Plus Protocol and Command Set* (Allen-Bradley, 1987).

2. Digital Equipment Corporation. *Digital Industrial Networks Guidebook* (Digital Equipment Corp., 1988).

3. Digital Equipment Corporation. *Networking: The Competitive Edge* (Digital Equipment Corp., 1985).

4. Seyer, M. D. *RS-232 Made Easy: Connecting Computers, Printers, Terminals, and Modems* (Prentice-Hall, 1984).

5. Stacy, A. H. *The Map Book: An Introduction to Industrial Networking* (Industrial Networking Inc., 1987).

6. Stone, H. S. *Microcomputer Interfacing* (Addison-Wesley Publishing, 1982).

7. Svacina, B. *Understanding Device Buses: A Tutorial* (Turck Inc., 1996).

8. Thiel, C. A., ed. *IBM Systems Journal Telecommunications*, vol. 18, no. 2 (IBM Corporation, 1979).

9. Thompson, L. M. *Industrial Data Communications Fundamentals and Applications*, 2nd ed. (ISA, 1997).

11

System Design and Applications

Introduction

The programmable controller-based control system offers a wide variety of configurations and capabilities. These range from a single machine or process control to an entire industrial plant control and monitoring system. After the decision has been made to use a programmable controller in a control application, the design engineer must complete the system design. This chapter will discuss a detailed system design method and then present several PLC applications.

System Design

The design of a programmable controller-based control system requires a simplified process flow diagram or equipment layout drawing, a process or machine control description, sizing and selection of the PLC equipment, a system specification, system drawings, wiring diagrams, and control programming. In this section, each of these areas will be discussed to provide a step-by-step approach to system design.

Piping and Instrument Drawings

The design of any process control system must start with the piping and instrument drawings (P&ID) or mechanical flow diagrams (MFD). These drawings show the process and/or mechanical equipment to be controlled and the instrumentation to be used in the control of the process or machine.

The process industry uses a standard set of symbols to prepare the P&IDs and MFDs. The symbols used in these drawings are generally based on the

ANSI/ISA S5.1-1984 (R1992), Instrumentation Symbols and Identification standard of the ISA as approved by the American National Standards Institute (ANSI). These drawings show how equipment is interconnected and what instrumentation is used to control the process. Figure 11-1 shows a typical P&ID.

In standard P&IDs, the process flow lines, such as process fluid flow and steam flow, are shown as heavy solid lines. The instrumentation signal lines are shown in a way that indicates whether they are pneumatic or electric. A cross-hatched line is used for pneumatic lines, for example, a 3-15 psi signal. The electric signal lines, usually 4-20 mA dc current, are normally represented by a dashed line.

A balloon symbol with an enclosed two- to four-letter code is used to represent the instruments associated with the process control loops. For example, in Figure 11-1 the balloon enclosing "TT-100" is a temperature transmitter and that enclosing "TIC-100" is a temperature-indicating controller. Generally, a number is assigned to each control loop. Combining the letter code and number into an instrument tag number labels the specific device in the loop, as illustrated in Figure 11-1.

Special items such as control valves and in-line instruments (for instance, orifice plates) have special symbols, as shown in Figure 11-1. Refer to ANSI/ISA S5.1 for a more detailed discussion of instrument symbols.

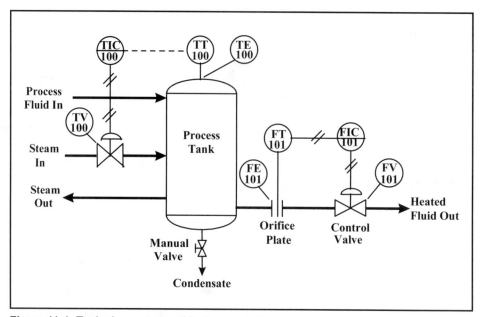

Figure 11-1. Typical process and instrument diagram.

The control system design engineer normally reduces the P&ID to a simplified process control diagram that shows only the equipment and instrumentation controlled or measured by the programmable controller. These simplified diagrams are then used to help programmers program the system by showing the status of the process in each step or state.

Process Description

The process description is probably the most important step in the design process because it conveys in simple language the purpose and the steps of the process. It is important because in most applications it is the main vehicle of communication between the user and the designer.

Sizing and Selecting a PLC System

Sizing a PLC system consists of estimating the number of input/output (I/O) modules required to control the process and the size of PLC memory needed. The selection process also consists of choosing the correct programming language and peripheral equipment for the control application.

I/O Sizing

A given PLC application may need many different types of I/O modules. Limit switches, pushbuttons, selector switches, motor controls, solenoids, and pilot lights may require either ac or dc modules of different voltage levels. Solid-state displays and some electronic instrumentation may require +5 vdc logic interface modules. Process instrumentation for measuring level, flow, temperature, or pressure may require analog-to-digital (A/D) conversion interfaces. Incremental encoders and stepping motors might need special-purpose I/O modules.

Interfacing these I/O devices and the PLC can be accomplished by using external electronic equipment to condition the signals that would otherwise increase the overall equipment cost for the system. Therefore, a PLC should be selected that has the correct I/O modules to match the type of field devices used in the process. This design process is called I/O sizing

Each programmable controller has a maximum number of input and output devices that can be monitored or controlled. Most PLC systems have I/O capacities ranging from a few to 8,096. These capacities can be divided into four PLC size categories: micro, small, medium, and large. Micro-PLCs normally have a limited number of built-in I/O circuits, 32 points or less. A small PLC system would range from 32 to 256 I/O points, a medium-sized PLC encompasses 256 to 1,024 I/O points, and a large PLC system has 1,024 and higher I/O points. To determine the PLC system size required, simply add up the number of field and control panel devices and compare the total to the above classifications. The system designer must also define the type and number of I/O because some PLC systems are constrained as to the mixture that can be interfaced to a given I/O system.

If modular I/O systems are being used, input and output point totals can then be used to determine the number and type of I/O modules required. Each type of module can interface a certain number of I/O, such as 4, 8, 16, or 32. Divide the number of inputs or outputs by the number of I/O points per module and round up to the nearest whole number. This calculation must be performed for each type of I/O module, i.e., discrete, pulse, analog, etc.

After defining the I/O requirements, the designer must also consider spares and future expansions. Most programmable controller users find that 10 percent to 20 percent spare capacity is enough for normal system growth.

Memory Sizing

The amount of memory required for an application is primarily a function of control program complexity and the number of I/O points in the system. The most precise method for determining memory size is to write out the control program and count the number of instructions used in the program. Then multiply this count by the number of words used per instruction. This number can be obtained from the PLC programming manual. Also consult the programming manual to see how much memory is used by executive programs and the processor overhead. The problem with this method is that, in a large system, the system programming sometimes takes months to complete, and the system must be designed and purchased in advance. A shortcut method is to multiply the number of I/O points by ten; this will yield a rough estimate of the memory required.

Example 11-1 illustrates memory sizing using this simpler method.

EXAMPLE 11-1

Problem: Estimate the memory size required for the PLC application shown in Figure 11-2.

Solution: To calculate the memory size, first calculate the number of I/O points in the system shown in Figure 11-2.

EXAMPLE 11-1 continued

Remote Area 1: I/O points = 70 + 35 + 6 = 111

Remote Area 2: I/O points = 95 + 50 + 10 = 155

Main Process Area: I/O points = 300 + 156 + 32 + 5 = 493

Total system I/O points = 111 + 155 + 493 = 759

Therefore, memory size = 10 x 759 = 7590 or 8K.

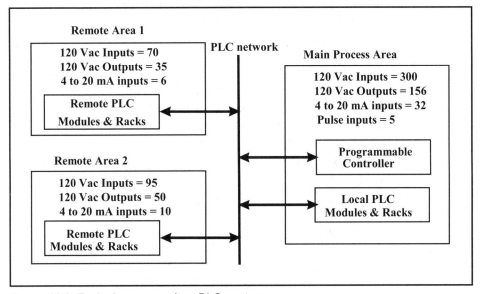

Figure 11-2. Typical process plant PLC system.

Selecting Programming Languages

As mentioned in chapter 10, the five basic types of programming languages available in programmable controller systems are ladder diagram (LAD), structured text (ST), statement list (STL), function block diagram (FBD), and sequential function chart (SFC). The type selected depends on the complexity of the control system and the technical background of the control system programmer and operators. Most PLCs offer the basic ladder logic instructions plus a combination of the other types of languages. The language most commonly selected is ladder diagram because this covers the basic ladder logic instructions and some data transfer and manipulations functions.

Peripheral Requirements

The term *peripheral* refers to the other equipment in the programmable controller system that is not directly connected to field I/O devices but increases the capabilities of the system. The most common peripheral is the programming device. This is generally available in three formats: a compact portable programming device from the PLC manufacturer, a laptop personal computer (PC) with PLC software installed, or a desktop personal computer that includes programming software. The compact portable programmer normally has a small, limited function keypad and a seven-segment LED display and can handle only one or two logic rungs at a time. It is normally used on micro or small PLC systems or for minor field changes to larger systems. The laptop PC-based programming device is normally used for field start-up and troubleshooting. The deskop personal computer is normally used in a lab or office environment to perform the development programming on PLC systems.

Another common peripheral used in PLC systems is a mass storage unit. This unit is used to store the control program on magnetic media so that, in the event of a program loss, the backup program can be reloaded into the controller memory. If a personal computer is used in the PLC system, the program can also be saved on a floppy disk or a hard disk for future use.

For hard-copy printouts of the control programs, a printer can be interfaced with the programming device, normally a PC, to obtain a program listing.

It is important for a complete system design that the peripheral equipment be available to back up and document the control program during start-up and field testing because reprogramming and documentation can be expensive and time consuming.

Other common peripherals are PROM programmers, process I/O simulators, and communications modules. The PROM programmers are used to write and save control programs on the PROM chips used in some controllers. The I/O process simulators are useful and cost-saving devices for large and complex systems that can be fully tested before installation and start-up in the field. The communications peripherals are used to communicate via communication networks between the programmable controllers and the plant or personal computers and other controllers in a system.

The type of operator interface to be used is one the most important considerations in a PLC system design. There are four main options for operator interfaces: (1) hard-wired local and main control panels; (2)

graphical user interface (GUI) software run on a personal computer; (3) intelligent peripheral devices, such as touch-screen operator interface; and (4) industrial PCs with function keys and GUI software. The system designer might also select a combination of several operator interface methods to implement a control system, such as hard-wired local panels and a GUI on a personal computer mounted in a central control room.

System Drawing and I/O Wiring Diagrams

A system drawing is used to give an overview of the system component interconnections and the communication cabling layout. This drawing is also useful to identify all the interface cables and components by model number. A typical PLC system drawing is shown in Figure 11-3. The system consists of a programmable controller, a personal computer with GUI graphics and PLC programming software installed, a printer with interface cable, a communications cable, a rack-mounted dc power supply, an eight-module I/O equipment rack, and the associated input/output modules.

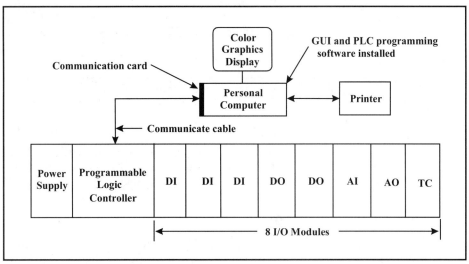

Figure 11-3. Typical PLC system diagram.

A typical discrete output module wiring diagram is shown in Figure 11-4. This drawing shows the wiring of process control equipment, such as heaters, pumps, and motors, to an ac output module. The wiring terminal strip in the programmable controller equipment cabinet is designated by TB-1, and a field junction box terminal strip is designated by JB-1. Field junction boxes are used extensively in process control applications to simplify field installation and instrumentation maintenance. Field wiring is normally indicated on wiring diagrams by a dashed line, as shown in

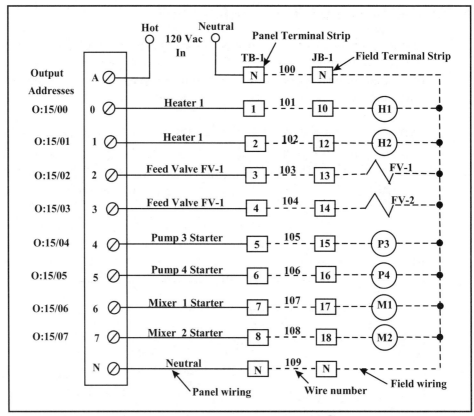

Figure 11-4. Typical discrete output module wiring diagram.

Figure 11-4. The wire numbers on the wiring diagram are used during the installation and maintenance of the system. The PLC output addresses are given on the left-hand side of the wiring diagram to help the engineer or technician start up or troubleshoot the control system.

System Programming

A PLC system can be programmed by the design engineer, plant operations personnel, maintenance, or the control system integrator. The choice of programming language should usually be left to plant operations or maintenance personnel since they will normally have to maintain the software after the system is installed.

We are now ready to discuss several PLC-based control system applications.

Natural Gas Dehydration Application

The first application to be discussed is the use of a PLC to control the dehydration process shown in Figure 11-5. The dehydration process uses small beads in the process tower to remove excessive moisture from natural gas.

In this process, the differential pressure switches (PSHL 1 and PSHL 2) are used to detect both high and low pressure across the dehydration tanks. The air-operated valves (AOVs) are used to control the gas flow to Towers 1 and 2. A typical diagram of the AOVs is shown in Figure 11-6. An electrically operated solenoid valve supplies instrument air at 80 psi to activate the control valves, and the valves have limit switches to indicate if they are open or closed.

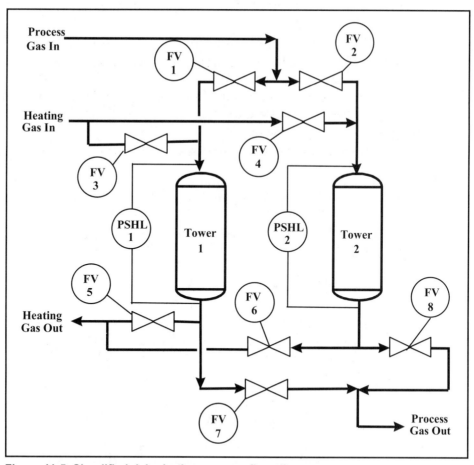

Figure 11-5. Simplified dehydration process flow diagram.

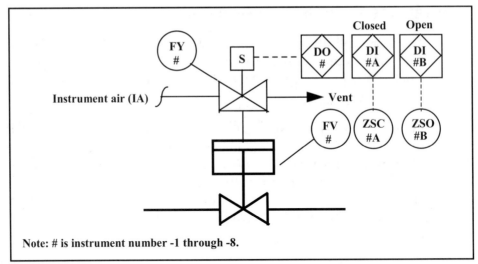

Figure 11-6. Detail for air-operated valve.

The programmable controller uses limit switches to determine the status or state of the valves. In some applications, only one limit switch is used, and the programmable controller can assume the opposite state (open or closed) if a single switch is used. However, a more reliable control is obtained if two switches are used. For example, if a valve fails to completely open or close on a given operation, the programmable controller is able to detect this failure and signal the operator.

The simplified process flow diagram of Figure 11-5 can be used to show the valve and/or equipment position for each state of the process.

The next step in the design process is to write a preliminary process control description.

Dehydration Process Control Description

The two process towers (towers 1 and 2) are used to remove moisture from natural gas. Generally, one tower is *in service* (i.e., removing moisture from the process gas) and the other tower is being dried out or *regenerated*. The automatic steps of the process are as follows:

1.   Assume that the control system has been placed in automatic, so the PLC can control the process based on the field inputs.

2.   If the differential pressure (dP) across tower 1 becomes high, as indicated by differential pressure switch (PSH-1), the programmable controller will place tower 1 in the regeneration mode by opening valves FV-3 and FV-5 and closing FV-1 and FV-7.

3.   At the same time, if the differential pressure across tower 2 is low, tower 2 will be placed in service by opening the gas valves FV-2 and FV-8, and control system will close the heating cycle valves FV- 4 and FV-6 for tower 2.

4.   Later, if the differential pressure across tower 2 becomes high and dP signal across tower 1 returns to low, the control system will place tower 1 in service and tower 2 into regeneration. This cycle will continue until the operator elects to stop the process.

After this preliminary control description is written, the designer is ready to size and select a programmable controller system.

Sizing and Selecting the PLC System

The first step is to decide which operator machine interface to use, such as local control panel or personal computer-based GUI. In this application, the dehydration process is a part of an entire natural gas processing plant, and plant operations personnel have decided to use a personal computer-based GUI for the operator interface. Figure 11-7 shows the graphics screen for tower 1 in service and tower 2 in regeneration for the dehydration process.

The operator uses the computer mouse to select the mode of operation required for the process. For example, if the pressure across tower 1 is low (i.e., the low-pressure [Low Press] box on the GUI display is highlighted), the operator can click on the In Service (In Serv.) button and place tower 1 in service to remove moisture from the natural gas.

The next step in the design process is to calculate the number of input and output modules required. This can be done by using the dehydration flow diagram in Figure 11-5 and the detail for the air-operated valves in Figure 11-6. There are eight (8) 120-vac solenoid valves, so we need eight 120-vac PLC outputs. There are 16 limit switches for the open and closed positions of the valves, and there are also four differential pressure switches. So, we need 20 discrete input points for this application. We can arbitrarily select 120 vac for the input signal voltage to save on the types of modules and supply voltages used in the system.

Assume that Allen-Bradley (A-B) 8-point, 120-vac I/O modules are used in the application. The next example shows how to calculate the number of I/O modules required.

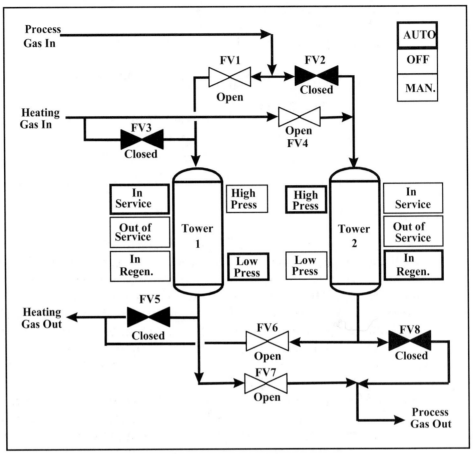

Figure 11-7. Dehydration GUI graphics display.

EXAMPLE 11-2

Problem: Estimate the number of A-B 8-I/O point, 120-vac modules required for the PLC dehydration application shown in Figure 11-5.

Solution: Calculate the number of I/O modules required for the dehydration system shown in Figure 11-5 as follows:

1) 120 vac input modules:

   (number of inputs + 20% x inputs)/ 8 inputs/module

   = (20 + 20 x .20)/8 modules

   = (24/8) modules

   = 3 input modules

EXAMPLE 11-2 continued

2) 120 vac output modules:

> (number of outputs + 20% x outputs)/ 8 outputs/module
>
> = (8 + 8 x .20)/8 modules
>
> = (9.6/8) modules
>
> = 2 output modules

System Drawing

The system drawing for this application will consist of an A-B PLC5/15 programmable controller, a personal computer installed with GUI and PLC programming software, a printer to document the program, and a single I/O rack with eight slots to hold the five I/O modules, as shown in Figure 11-8. The system diagram is normally plotted on a "D" size (24 inches by 36 inches) drawing sheet and would include a detailed material list. This list includes the equipment number, material description, manufacturer, and model number, as shown in Table 11-1.

For this application, an Allen-Bradley PLC5-15 has been selected for the programmable controller and an IBM Pentium personal computer has been selected for the GUI and PLC programming software. Also, an

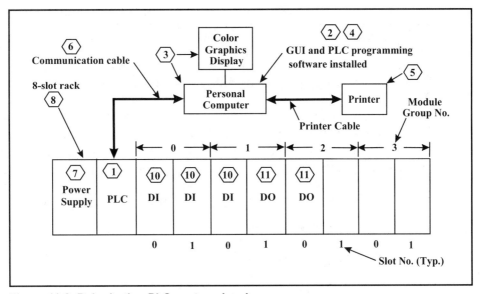

Figure 11-8. Dehydration PLC system drawing.

eight-slot A-B chassis was chosen to hold the AC input and output modules, as listed in Table 11-1.

The placement of the input modules and output modules in an I/O equipment rack is important. For example if the system has a mix of ac, dc, and analog modules, the ac modules are normally segregated from low-level signal modules, such as analog I/O (4 to 20 mA dc), 5 vdc inputs, and millivolt (mV) input modules, to avoid electrical interference.

Table 11-1. Dehydration Control System Material List

Equip. No.	Material Description	Manufacturer	QTY	Model No.
1	Programmable controller	Allen-Bradley	1	1785-LT.
2	PLC programming software	Allen-Bradley	1	6200
3	Personal computer	IBM	1	Pentium
4	GUI software	Rockwell	1	RS-View
5	Graphics printer	Hewlett-Packard	1	LaserJet5
6	Communication cable	Allen-Bradley	1	1784-CP.
7	PLC5 rack power supply	Allen-Bradley	1	1775-P1
8	8-slot PLC chassis	Allen-Bradley	1	1771-APB
9	120 vac input module	Allen-Bradley	3	1771-IA
10	120 vac output module	Allen-Bradley	2	1771-OA

I/O Wiring Diagrams

The I/O wiring diagram for the first ac output module is shown in Figure 11-9. This drawing shows the wiring of the system solenoid valves to the ac output module. In the sample application, a single ac line is used where the ac hot is normally designated by L1 and ac neutral has wire number L2. However, in larger systems more than one ac line might be used, depending on loading requirements. The design engineer must also consider maintenance of the system, so in the sample application the wiring for tower 1 was connected to one ac power circuit and the wiring to the other tower was connected to a second ac power circuit. In this case, one tower could be taken out of service, and the other tower could be kept on line during maintenance of the second tower. Furthermore, the wiring for the control panel might be placed on a third ac power circuit.

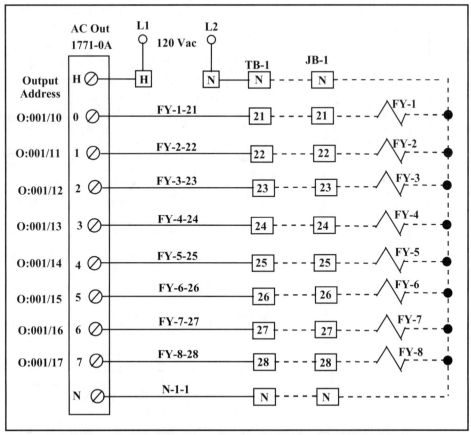

Figure 11-9. Wiring diagram for solenoid valves on dehydration system.

The wiring of the ac input modules is performed in a similar manner, except that the field inputs are normally drawn on the left side of the input module as shown in Figure 11-10. This input wiring diagram shows how to connect the high- and low-pressure switches for towers 1 and 2 to the first AC input module. As the system drawing shows, the first ac input module is in rack 00, module group 0, slot 0, so the Allen-Bradley input addresses are between bit I:000/00 and bit I:000/07, as shown on the right-hand side of the module input drawing.

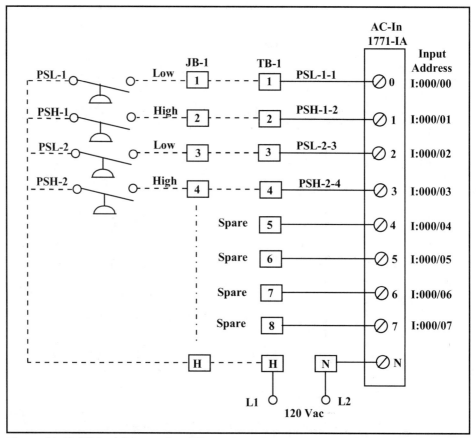

Figure 11-10. Wiring diagram for differential pressure switches.

The input wiring diagram in Figure 11-11 shows the connection of the valve limit switches for control valves FV-1 through FV-4. As the system drawing shows, this ac input module is in rack 00, module group 0, slot 1. The input addresses are therefore between bit 111/10 and bit 111/17, as shown on the right-hand side of the module input drawing.

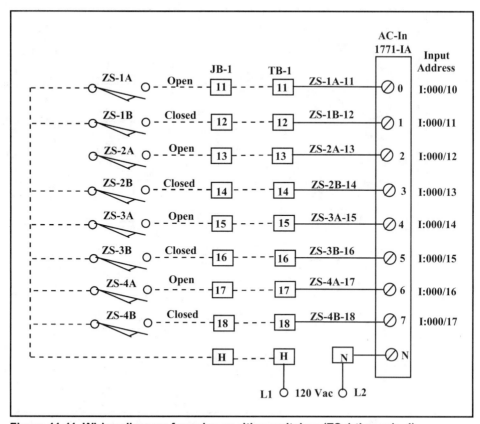

Figure 11-11. Wiring diagram for valve position switches (ZS-1 through -4).

The wiring diagram for the limit switches on the remaining four control valves (i.e., FV-5, FV-6, FV-7, and FV-8) is shown in Figure 11-12. This ac input module is in rack 00, module group 1, so the input addresses are between bit I:001/00 and bit I:001/07, as shown on the right-hand side of the module input drawing.

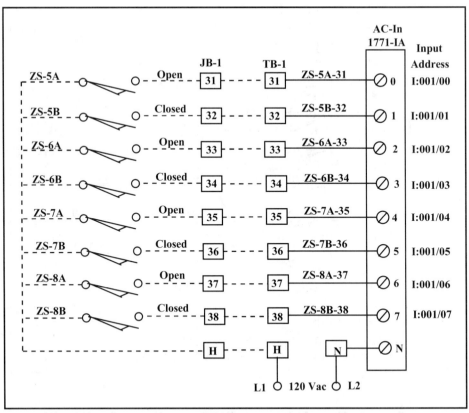

Figure 11-12. Wiring diagram for valve position switches (ZS-5 through -8).

Application Programming

In a small system like the dehydration application in our example, basic ladder programming is the best choice. This is because the system requires only simple on/off control, and there are no involved analog or data manipulations anticipated. The first step in the logic programming is to make a list of internal bits that interface with the GUI. The next step is to list the I/O points in the system using the I/O module wiring diagrams. The internal bits that interface with the GUI are listed in Table 11-2.

Table 11-2. GUI Bit Assignments

Bit Address	Description	Bit Address	Description
B3:0/00	Automatic mode	B3:0/05	T1In regeneration
B3:0/01	Off	B3:0/06	T2In service
B3:0/02	Manual mode	B3:0/07	T2Out of service
B3:0/03	T1in service	B3:0/08	T1In regeneration
B3:0/04	T1Out of service		

These bit assignment tables are an aid in programming because they consolidate the I/O information in a single table for easy reference. In most PLC programming software packages, these lists can be entered directly into the control program using the programming software.

This ladder program operates in a relatively simple manner. When the operator clicks on the AUTO button shown in the upper right-hand corner of Figure 11-7, this sets internal bit B3/00 to logic 1 and places the control system into the automatic mode. The GUI software is programmed to deactivate the OFF and manual (MAN.) buttons on the screen, if the operator selects the AUTO mode.

The ladder logic for the automatic mode of the process can be written, as shown in Figure 11-13. Tower 1 is placed in service if the differential pressure in the tower is low, the AUTO function (internal bit B3:0/00) is TRUE, and the tower 1 differential pressure is not high (bit I:000/00). The output for "tower 1 in service" is sealed in with its own control bit B3:0/3. Tower 1 will stay "in service" until its differential pressure reaches a high level as measured by a high differential pressure (PSH-1) across the tower. If the dP switch PSH-1 is closed, then input bit I:000/00 is set to TRUE, and its normally closed contacts are opened. So, the "tower 1 in service" output coil bit B3/03 is turned OFF.

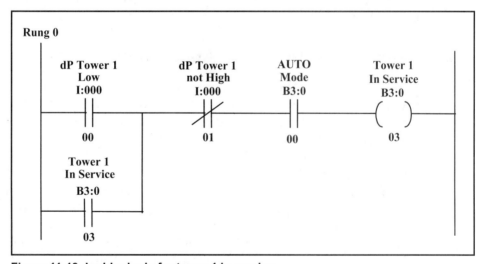

Figure 11-13. Ladder logic for tower 1 in service.

Using the ladder logic software shown in Figure 11-14, tower 1 is placed into regeneration by the control system to remove the moisture that has built up during the service cycle. The valves are designed to fail closed (FC), so if plant air or electric power is lost, the valve will close. Therefore, the solenoid valves must be energized to open the valve. The ladder logic

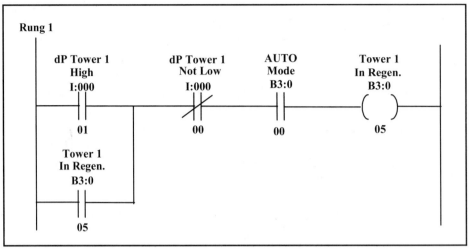

Figure 11-14. Ladder logic for tower 1 in regeneration.

program in Figure 11-15 shows the control logic for solenoid valves FY-1, FY-7, FY-3, and FY-5. The same design procedure can be used to write the control software for tower 2.

The valve limit switches are used to animate process graphics valve symbols during each step in the process. If a valve is closed, the valve symbol will be filled in by the GUI software using the information sent to the PC from the PLC.

Two-Stage Alternating Pump Application

In this application, the ladder logic program to alternate pumps in processes like the emptying of wells, reservoirs, and tanks will be developed. In this type of application, two smaller pumps are frequently used instead of one large one to reduce cost.

Alternating pump operation (pump 1 as the primary, then pump 2 as the primary) reduces the maintenance required on the individual pumps and provides more reliable operation. In this application, the secondary or "standby" pump is available if the rate of water entering the vessel is more than the first pump can handle. If this situation occurs, the second pump will also turn On and assist the primary pump. The triggers for these events could be analog signals or simple discrete inputs (level switches, etc.).

The first step in the design of the PLC-based control system is to draw the piping and instrument diagram as shown in Figure 11-16.

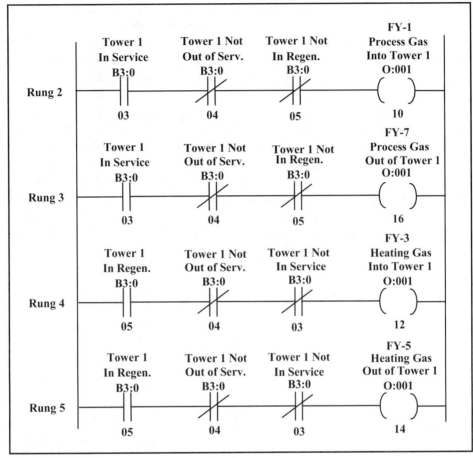

Figure 11-15. Ladder logic for tower 1 valves.

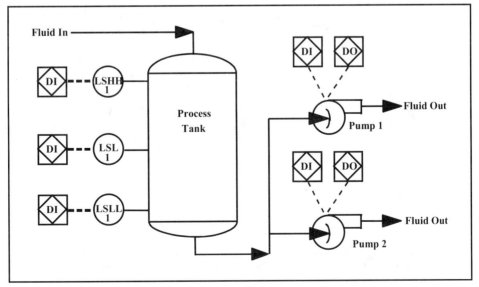

Figure 11-16. Alternating pump control of tank level.

The next step in the design procedure is to select a PLC and I/O types. Since there are only five discrete inputs and two discrete outputs, a micro-PLC is sufficient for this application. An A-B Micro 1000 PLC with ten discrete inputs and six discrete outputs was selected for this application. The design of the input/output wiring is shown in Figure 11-17.

The ten inputs (addresses I:0/0 through I:0/9) are shown on the left side of the micro-PLC, and the five AC outputs (addresses O:0/0 through O:0/5) are drawn on the right side of the unit. In this diagram, three normally open (NO) level switches, LSHH-1, LSL-1, and LSLL-1, were connected to the first three inputs on the micro-PLC at inputs I:0/0, I:0/1 and I:0/2, respectively. Two normally opened auxiliary (aux) contacts from the pump motor starters (M1 and M2) were connected to inputs I:0/4 and I:0/5. The pump motor starter relays are connected to the first two AC outputs on the micro-PLC at points O:0/0 and O:0/1.

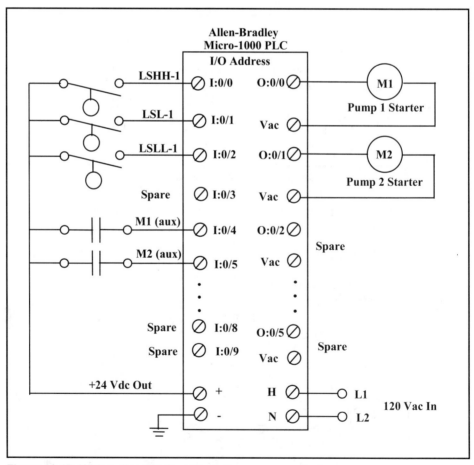

Figure 11-17. Wiring diagram for alternating pump control.

The final design step is to write the control program for this application. The ladder logic used in this application consists of four rungs. Rungs 0 and 1 form an alternating circuit, so each time the fluid in the tank is low, switch LSLL-1 is closed. This sets bit I:0/2 to 1, and the alternator bit in rung 1, B3:0/2, changes state. The status of this bit determines which pump will turn on first. The one-shot rising (OSR) instruction in rung 0 is a specialized instruction that is only energized for one processor scan. This causes internal bit B3:0/1 to be energized for one processor scan if the low-low level switch (bit I:0/2) is closed as shown in rung 0 of Figure 11-18.

Rung 2 controls the operation of pump 1, using output bit O:0/0 on the micro-PLC. If the low-low level switch is closed, it sets bit I:0/2 to 1. If the alternator internal bit B3:0/2 is turned off, and if the level in the tank has reached the low-level switch (LSL-1), setting bit I:0/1 to 1, then pump 1 will be the first pump to be energized. If the alternator bit B3:0/2 is on, then pump 1 will be the second pump to be energized, as shown in rung 2 of Figure 11-19.

Rung 3 controls the operation of pump 2 using output bit O:0/1. If the low-low-level switch is on (bit I:0/2 is set), the alternator bit B3:0/2 is on, and the level in the tank has reached the low-level switch (i.e., Bit I:0/1 is set to 1), then pump 2 will be the first pump to be energized. If B3:0/2 is off, pump 2 will be the second pump to be energized, as shown in rung 3 of Figure 11-20.

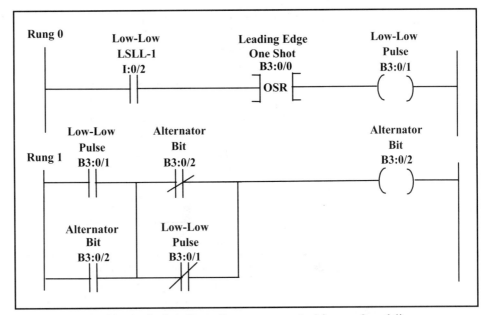

Figure 11-18. Ladder logic for alternating pump control (rungs 0 and 1).

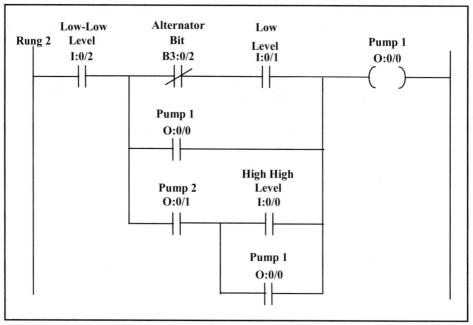

Figure 11-19. Ladder logic for alternating pump control (rung 2).

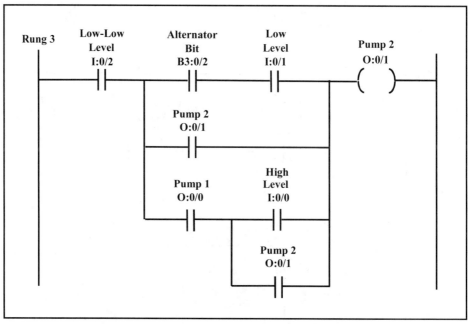

Figure 11-20. Ladder logic for alternating pump control (rung 3).

EXERCISES

11.1   List the main items required to design PLC systems.

11.2   What is the function of piping and instrument drawings?

11.3   What is the purpose of process description in PLC system design?

11.4   Estimate the memory size required for a PLC application with the following I/O points: (1) Main Control Room: 150 points, (2) Remote Process Area 1: 254 points, (3) Remote Process Area 2: 125 points, and (4) Remote Process Area 3: 156 points.

11.5   List some of the peripheral equipment commonly used in PLC applications.

11.6   What is the purpose of a PLC system drawing?

11.7   Calculate the number of input and output modules required on the dehydration system in the example used in this chapter, if a third process tower is added. Assume that all inputs and outputs are 120 vac.

11.8   Write a ladder logic program to control the valves on tower 2 of the dehydration application shown in Figure 11-5.

11.9   Revise the alternating pump application LAD program to use the pump starter auxiliary (AUX.) contact inputs instead of the pump starter output bits.

BIBLIOGRAPHY

1.   Allen-Bradley Co., Inc. *Allen-Bradley Industrial Computer and Communications Group Product Guide* (Allen-Bradley, 1987).

2.   Allen-Bradley Co., Inc. *Allen-Bradley Programmable Controller Products* (Allen-Bradley, 1987).

3.   Allen-Bradley Co., Inc. *Micro Mentor: Understanding and Applying Micro Programmable Controllers* (Allen Bradley, 1995).

4.   Allen-Bradley Co., Inc. *Processor Manual PLC-5 Family Programmable Controllers* (Allen-Bradley, 1987).

5.   Jones, C. T., and L. A. Bryan. *Programmable Controllers: Concepts and Applications* (International Programmable Controls, Inc., 1983).

12

Design, Installation, and Maintenance

Introduction

To complete the discussion of programmable controller system concepts and design principles, we need to cover panel design, equipment installation and layout, operational testing, maintenance, troubleshooting, and documentation. The reliability and maintainability of a control system is, in a large part, a function of proper system design and documentation that considers the maintenance aspects of a system.

Control Panel Design

The main feature that separates PLCs from other types of computers is that programmable controllers are designed to be installed in harsh industrial environments. They are, in some cases, simply installed on metal sheet (subpanel) on the production floor. However, in most cases, programmable controller system components are installed in a metal enclosure or a control panel to protect them against atmospheric contaminants such as dust, moisture, oils, and other corrosive airborne substances. These metal enclosures also reduce the effects of electromagnetic radiation generated by electrical or welding equipment.

The panel or enclosure design requires a panel layout design, considerations for heating and maintenance, wiring layout and duct design, power distribution design, and normally the writing of a panel specification.

Control Panel Layout

The panel size depends on the amount of equipment to be installed in the enclosure and whether front panel controls and instruments are required on the system. If front panel controls are required, the size and shape of the panel are mainly controlled by the best layout design of these front panel instruments. To assure correct and easy operation of the system, the indicators and recorders are placed at normal eye level and hand switches are placed below the indicators.

The metal enclosure should conform to industrial standards, such as the standard issued by the National Electrical Manufacturers Association (NEMA). The NEMA standard covers the design of industrial enclosures for different industrial environments. The equipment layout inside the panel should follow the recommendations contained in the installation manual for your programmable controller. In this manual, the manufacturer generally lists the minimum spacing allowed between I/O racks, other equipment, and the processor. This spacing generally takes into consideration equipment heating, electrical noise, and safety factors. Figure 12-1 shows a typical minimum equipment spacing of six inches for a programmable controller system based on a PLC manufacturer's recommendation.

Heating Considerations

To allow for effective convection cooling, most manufacturers recommend that all system components be mounted in a position that maximizes airflow in the enclosure. Since the power supplies generate the most heat, they should not be mounted directly underneath another system component. Generally, the main power supply is mounted near the top of the enclosure, but some PLC manufacturers use individual power supplies on each I/O rack and an internal power supply for the processor. In Figure 12-1, notice the six-inch horizontal gap between the I/O racks, which allows for an adequate level of cooling airflow.

The temperature inside the control panel or enclosure must not exceed the maximum operating temperature listed in the manufacturer's installation manual (typically 120°F). If that temperature limit cannot be maintained by convection cooling, a fan or blower must be installed to help dissipate the heat. The fan or fans you use will generally be equipped with filters to prevent dust, dirt, and other airborne contaminants from entering the enclosure and affecting the system components.

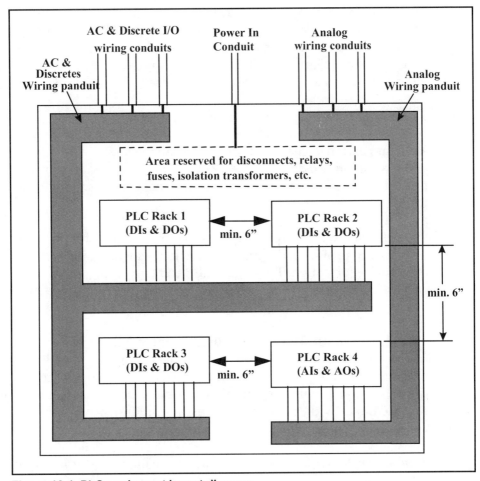

Figure 12-1. PLC equipment layout diagram.

Enclosure Standards

All enclosures installed in industrial applications must meet the NEMA standard described in NEMA publication number 250-1979. The following descriptions are excerpts from this NEMA standard, and the enclosure types are those rated for ``Nonclassified Locations'' and ``Classified Locations'' (i.e., explosive locations).

Nonclassified Location Enclosures Types

Type 1 Enclosures. Type 1 enclosures are intended for indoor use, primarily to provide a degree of protection against contact with the enclosed equipment in locations where unusual service conditions do not exist. The enclosures shall meet the rod entry and rust-resistance design tests.

Type 2 Enclosures. Type 2 enclosures are intended for indoor use, primarily to provide a degree of protection against limited amounts of falling water and dirt. These enclosures shall meet rod entry, drip, and rust-resistant design tests. They are not intended to provide protection against conditions such as internal condensation or internal icing.

Type 3 Enclosures. Type 3 enclosures are intended for outdoor use, primarily to provide a degree of protection against windblown dust, rain, sleet, and external ice formation. They shall meet rain, external icing, dust, and rust-resistance design tests. They are not intended to provide protection against conditions such as internal condensation or internal icing.

Type 3R Enclosures. Type 3R enclosures are intended for outdoor use, primarily to provide a degree of protection against falling rain, sleet, and external ice formation. They shall meet rod entry, rain, external icing, and rust-resistance design tests. They are not intended to provide protection against conditions such as dust, internal condensation, or internal icing.

Type 4 Enclosures. Type 4 enclosures are intended for indoor or outdoor use, primarily to provide a degree of protection against windblown dust and rain, splashing water, and hose-directed water. They shall meet hose-down, dust, external icing, and rust-resistance design tests. They are not intended to provide protection against conditions such as internal condensation or internal icing.

Type 4X Enclosures. Type 4X enclosures are intended for indoor or outdoor use, primarily to provide a degree of protection against corrosion, windblown dust and rain, splashing water, and hose-directed water. They shall meet the hose-down, dust, external icing, and corrosion-resistance design tests. They are not intended to provide protection against conditions such as internal condensation or internal icing.

Type 5 Enclosures. Type 5 enclosures are intended for indoor use, primarily to provide a degree of protection against dust and falling dirt. They shall meet the dust and rust-resistance design tests. They are not intended to provide protection against conditions such as internal condensation.

Type 6 Enclosures. Type 6 enclosures are intended for indoor or outdoor use, primarily to provide a degree of protection against the entry of water during occasional temporary submersion at a limited depth. They shall meet submersion, external icing, and rust-resistance design tests. They are not intended to provide protection against conditions such as internal condensation, internal icing, or corrosive environments.

Type 6P Enclosures. Type 6P enclosures are intended for indoor or outdoor use, primarily to provide a degree of protection against the entry of water during prolonged submersion at a limited depth. They shall meet air pressure, external icing, and corrosion-resistance design tests. They are not intended to provide protection against conditions such as internal condensation or internal icing.

Type 11 Enclosures. Type 11 enclosures are intended for indoor use, primarily to provide, by oil immersion, a degree of protection to enclosed equipment against the corrosive effects of liquids and gases. They shall meet drip and corrosion-resistance design tests. They are not intended to provide protection against conditions such as internal condensation or internal icing.

Type 12 Enclosures. Type 12 enclosures are intended for indoor use, primarily to provide a degree of protection against dust, falling dirt, and dripping, noncorrosive liquids. They shall meet drip, dust, and rust-resistance tests. They are not intended to provide protection against conditions such as internal condensation.

Type 12K Enclosures. Type 12K enclosures with knockouts are intended for indoor use, primarily to provide a degree of protection against dust, falling dirt, and dripping, noncorrosive liquids other than at knockouts. They shall meet drip, dust, and rust-resistance design tests. Knockouts are provided in the top and/or bottom walls only. After installation, the knockout areas shall meet the environmental characteristics listed earlier. They are not intended to provide protection against conditions such as internal condensation.

Type 13 Enclosures. Type 13 enclosures are intended for indoor use, primarily to provide a degree of protection against dust, spraying water, oil, and noncorrosive coolant. They shall meet oil exclusion and rust-resistance design tests. They are not intended to provide protection against conditions such as internal condensation.

Classified Location Enclosures

Type 7 Enclosures. Type 7 enclosures are for indoor use in locations classified as Class I, Groups A, B, C, or D, as defined in the *National Electrical Code.*

Type 8 Enclosures. Type 8 enclosures are for indoor or outdoor use in locations classified as Class II, Groups A, B, C, or D, as defined in the *National Electrical Code.*

Type 9 Enclosures. Type 9 enclosures are intended for indoor use in locations classified as Class II, Groups E, F, or G, as defined in the *National Electrical Code*.

Type 10 Enclosures (MSHA). Type 10 enclosures shall be capable of meeting the requirements of the Mine Safety and Health Administration, 30 C.F.R., Part 18 (1978).

Maintenance Design Features

The system designer must include certain features in the design of the enclosure to reduce maintenance time and cost. One important consideration is the accessibility of equipment components and terminal connections. For example, the processor should be placed at a normal working level for ease of operation or maintenance. If the processor and the system power supply are contained in a single unit, it should be placed near the top of the enclosure. However, if there is adequate space in the enclosure, it should be placed to improve operation or maintenance.

Another maintenance feature is that the control panel should have an ac power outlet strip so maintenance can plug in test equipment and a portable light if needed. If the panel is large, interior lighting should be installed to aid maintenance and operations personnel in troubleshooting. Both the ac power strip and the interior lighting should be on a separate ac circuit from the other system components.

In some harsh environmental applications, a gasketed glass window is used to allow operations and maintenance personnel to view the processor status lights and/or I/O status indicators without opening the panel door. This is an important feature in dirty and corrosive environments because it prevents damage to the control components in the panel. In some applications, the operator will check the status of process and I/O lights during each step in a process to make sure there are no abnormal operating conditions.

Panel Duct and Wiring Design

The wiring duct layout is determined by the types of signals used in the system design. It also depends on where the I/O modules are placed in the racks. The main consideration is reducing the electrical noise caused by cross talk between I/O signal lines.

All ac power wiring should be kept separated from low level dc wires. If dc lines must cross ac signal or power lines, it should be done at right angles only. This routing practice minimizes the possibility of electrical interference.

Power Distribution Design

Most programmable controller manufacturers recommend that a power isolation transformer be installed between the ac power source and the PLC equipment to provide for signal isolation from other equipment in the process area. A typical power distribution drawing is shown in Figure 12-2. In this application, the incoming ac lines (L1, L2, and L3) are 460 vac, which is used for power devices, such as motors, heaters, or pump starters in the system. There are three fuses on the incoming lines to protect against any over-current condition. In the figure, the 460 vac power lines L1 and L2 are connected to the primary of the step-down transformer. This transformer "steps down" or reduces the ac voltage to 120 volts and isolates the programmable controller's components from the outside electrical equipment.

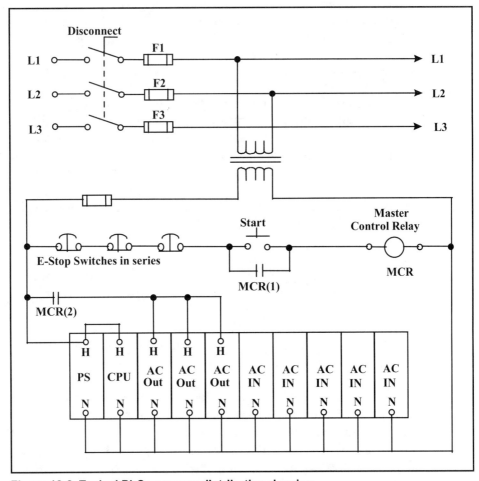

Figure 12-2. Typical PLC ac power distribution drawing.

The ac distribution circuit contains a ``master control relay,'' labeled as MCR in the circuit diagram. This relay is used to stop the programmable controller or the controlled machine or process when an operator depresses any emergency stop (E-stop) pushbutton. Any number of E-stop switches can be used in the ac distribution circuit to improve system safety.

It is also recommended that emergency stop circuits be designed into the system for every machine being directly controlled by the programmable controller. These circuits should be hard-wired and totally independent of the programmable controller to ensure maximum safety in the control system.

Programmable controllers are very reliable devices, but if the central processor fails, it can cause the control system to behave dangerously and erratically. Therefore, the operator must be able to quickly and safely turn off process equipment and machinery. This can be done by placing E-stop switches in locations that are easily accessible to the operator.

Grounding Considerations

The reliability of any electrical control system is highly dependent on the proper design of system grounding. Correct electrical grounding design is also required to guarantee a safe electrical installation. When designing and installing any electrical equipment in the United States, the designer should read and understand Section 250 of the National Electrical Code (NEC). This section provides information on the connection methods and wire color code, size, and type of conductor required to ensure that electrical equipment is grounded safely.

The code states that a ground must be permanent (i.e., no solder connections), continuous, and able to safely conduct the current in the system with minimal resistance. The ac hot wire must have black-colored insulation, the ac neutral wire insulation must be white, and the ground conductors must use green-colored insulation. It is common practice to use red insulation for positive signal wires and black-colored wires for negative dc signal lines, but it is not required by NEC.

The ground wire should be separated from the ac hot and ac neutral wires at the point of entry to the control panel. To minimize the ground wire length within the panel, the ground reference point should be located as close as possible to the point of entry of the panel supply line. All I/O racks, power supplies, processors, and other electrical devices in the system should be connected to a single ground bus in the control panel. Paint or other nonconductive materials should be scraped away from the area where an I/O rack makes contact with the panel. In addition to the ground connection made through the rack mounting bolts, a metal braid of the size recommended by the PLC manufacturer should be used to connect each chassis and the panel at a single mounting bolt.

I/O Module Installation and Wiring

An important part of the hardware design of a PLC system involves the proper layout and wiring diagrams for the input/output module. The actual installation of the modules is a relatively simple procedure with most of the programmable controllers on the market, since the installation crew simply plugs the modules into the I/O racks per the drawing provided by the system designer.

However, in some cases, intelligent modules (such as thermocouple, analog, communications, etc.) might require switch settings to properly configure the module. For example, a thermocouple (T/C) module might be designed to accept the various T/C types, such as J, K, T, S, etc., and the user sets toggle switches in the module to accept the T/C being used in the process. The design engineer must document the switch setting on the I/O drawings or in the installation instructions to guarantee that the modules are properly installed and configured.

Another important procedure required for installation of a programmable controller system is the selection of the I/O chassis number for each I/O rack. Many PLC manufacturers use switch assemblies in the I/O rack to number the racks. A typical programmable controller I/O rack selection switch arrangement is shown in Table 12-1.

Table 12-1. Typical I/O Rack Number Selection Table

I/O Rack Number	Switch Position			
	4	3	2	1
00	Closed	Closed	Closed	Closed
01	Closed	Closed	Closed	Open
02	Closed	Closed	Open	Closed
03	Closed	Closed	Open	Open
04	Closed	Open	Closed	Closed
05	Closed	Open	Closed	Open
06	Closed	Open	Open	Closed

Control Panel Specification

In most cases, the equipment enclosure for a programmable controller system is fabricated by an outside control panel vendor so that an equipment specification will be required to cover all requirements of a system. The following is a typical control panel specification:

The control panel furnished under this specification shall be supplied complete with the instruments and equipment listed on the enclosed drawings, installed and electrically wired, and ready for wiring to field instruments and equipment.

The control panel shall conform to the following:

1.   The control panel shall be fabricated with cold-rolled steel plate.

2.   All miscellaneous items, such as wire raceways, terminal strips, electrical wire, and so on, where possible shall be made of fire-resistant materials.

3.   The grounding bus bar and studs shall be pure copper metal.

4.   A minimum of two 120-vac, 60-Hz utility outlets shall be installed in the panel.

5.   An internal fluorescent light with a conveniently located ON/OFF switch shall be installed in the panel.

6.   A circuit breaker panel shall be provided that has circuit breakers to accommodate all panel lighting, power supplies, instruments, I/O modules, processors, and any other loads listed on the system drawings.

7.   Each electric wire over 12 inches in length shall be identified at each end with a wire number per the electrical drawings for the system.

8.   Terminal strips with screw-type connectors shall be used, and no more than two wires shall be terminated on any single terminal.

9.   Nameplates shall be made of laminated plastic with white letters engraved on a black background for both front and rear panel-mounted instruments and components, such as power supplies, transformers, and PLC I/O racks.

10.   The ac wiring shall be separated from 4-to-20-mA dc current and digital signals by a minimum of 24 inches, and the ac wiring and dc signals must be wired to separate terminal strips.

11.   All electrical wires and cables shall enter the control panel through the top. Sufficient space shall be provided to allow ac power cables entering the top of the panel to continue directly to the circuit breaker panel.

12.   All wires and cables shall be routed and tied to provide clear access to all instruments and components for maintenance purposes and for the removal of defective components.

13.   All dc wires shall have a minimum insulation voltage rating of 600 volts ac, and all dc wires shall have a minimum insulation voltage rating of 300 volts dc.

14.   All electrical conductors shall be copper of the correct wire size for the current carried with 98 percent conductivity, referenced to pure copper.

15.   Wireways shall be attached securely to the control panel.

16.   Metal surfaces that have wires or cables passing through them shall be furnished with insulated polyethylene grommets to prevent damage to the conductors or cables.

17.   All wire bundles or cables shall be clamped to the panel at all right-angle turns.

18.   All wires entering or leaving a wire bundle shall be tied to the bundle at the point at which they enter or leave the main wire bundle.

Equipment Layout Design

Proper control system equipment layout design can reduce installation costs and improve system reliability and maintainability. In addition to the programmable controller components, the equipment layout must also take into account the other system components, such as field devices and instruments, power disconnect boxes, power transformer, and the location of process equipment and machines.

In general, placing the processor near the process equipment and using remote I/O racks where possible will reduce wire and electrical conduit runs. It is possible that the cost of wiring and conduit installation will far exceed the programmable controller equipment costs if care is not taken in the equipment layout design.

The control panel should be placed in a position that allows the doors to be opened fully for easy access to wiring terminals and system components for maintenance and troubleshooting. The National Electric Code (NEC) requires that the panel doors open at least to 90°, and there must be a minimum of 36 inches of clearance from the rear of the panel to the nearest grounding surface or wall. Before starting the design of a control panel, the designers should read and understand section 110 of the National Electrical Code to avoid any code violations or safety problems.

An emergency disconnect switch should be mounted on or near the control panel in an easily accessible location. If the location where the control panel will be placed contains equipment that generates excessive radio frequency interference (RFI) or electromagnetic interference (EMI), the panel should be placed a reasonable distance from these sources. Examples of such sources include electric machinery, welding equipment, induction heating units, and electric motor starters.

System Start-Up and Testing

The first requirement of a successful system start-up is to have a written system operational (SO) test procedure. This procedure should be written by the control system designer and carefully reviewed and approved by all parties involved in the project. To save time, use a process simulator to functionally test the control system program before the system is installed on the process.

The SO test procedures will generally contain the following main sections: visual inspection, continuity test, testing input signals, testing outputs, and testing the process operation. A typical SO test procedure is described in the following sections.

Typical SO Test Procedure

I. Visual Inspection

1. Verify that all system components are installed according to the system drawing.

2. Check that the I/O module is located in the equipment racks according to the I/O drawings.

3. Inspect that the switch settings on all intelligent modules conform to the drawings.

4. Verify that all system communication cables are correctly installed.

5. Check that all input wires are correctly marked with wire numbers and terminated at the correct points on the input modules.

6. Check that all output wires are correctly marked with wire numbers and are terminated at the correct terminals on the output modules.

7. Verify that the power wiring is installed according to the ac distribution drawing.

II. Continuity Check

1. Use an ohmmeter to verify that no ac wire is shorted to ground.

2. Verify the continuity of the ac hot wiring.

3. Check the ac neutral wiring system.

4. Check the continuity of the system grounds.

III. Input Wiring Check

1. Place the programmable controller in the program mode.

2. Disable all output signals.

3. Turn ``on'' the ac system power and the power to the input modules.

4. Verify that the E-stop switch removes power from the system.

5. Activate each input device, observe the corresponding address on the programming terminal, and verify that the indicator light on the input module is energized.

IV. Output Wiring Check

1. Disconnect all output devices that might create a safety problem, such as motors, heaters, control valves, and the like.

2. Place the programmable controller in "Run/Program" mode.

3. Apply power to the programmable controller and the output modules.

4. Depress the E-stop pushbutton, and verify that all output signals are deenergized.

5. Restart the system and use the forcing function in the programming terminal to energize each output individually.

6. Measure the signal at the output devices and verify that the output light on the module is energized for each output you test.

V. Operation Test

1. Place the processor in the "Run/Program" mode and turn on the main power switch.

2. Load the pretested control program into the programmable controller.

3. Disable all outputs, select the "Run/Program" mode on the processor, and verify that the run light on the processor is activated.

4.   Check that each rung of logic is operating properly by simulating the inputs and by verifying on the programming terminal that the correct output is energized at the proper time or sequence in the program.

5.   Make any required changes to the control program.

6.   Enable the output modules, and place the processor in ``Run'' mode.

7.   Test that the control system functions according to the process operating procedure.

Maintenance Practices

Programmable controller components are designed to be very reliable, but occasional repair is still required. System maintenance costs can be greatly reduced by using good design practices, complete documentation, and preventive maintenance programs.

Preventive Maintenance

A systematic preventive maintenance program will reduce the down time of the control system. The preventive maintenance of programmable controller components is usually scheduled at the same time as the machine or process that is down for maintenance or repair. Normally, since programmable controller equipment is more reliable than some machinery or process equipment, it requires less frequent preventive maintenance operations.

It is very important in a preventive maintenance program to carefully check the various connectors in the system. It is estimated that 70 percent of the problems found in electronic or computer-based systems are caused by loose, dirty, or defective connectors or connections. The connections to the I/O modules should be checked periodically to make sure no wiring has come loose. The seating of the I/O modules in the equipment rack should also be checked. If the system is located in process areas that have high vibration levels, the preventive maintenance check should be performed more often.

Excessive heat is another major cause of failure in programmable controller systems. Therefore, if an enclosure is cooled with fans, the filters used on the panel must be cleaned or replaced on a regular basis. It is important that dirt and dust is not allowed to build up on system components because a dirt buildup on electronic components can reduce heat dissipation and cause an overheated condition in the system.

Electrical noise can cause erratic and dangerous operation of a PLC system, so the maintenance department must check to make sure that equipment producing high levels of RFI or EMI noise are not moved near the programmable controller equipment.

Maintenance personnel should also check to make sure that unnecessary items are not stored on or near the PLC equipment. Leaving items such as tools, test equipment, drawings, or instruction manuals in the control panel can obstruct the air flow and cause heating problems.

Recommended Maintenance Measures

To reduce the risk of accidents and improve ease of maintenance, it is recommended that the following measures be taken: 1) stocking of critical spare parts, 2) documentation of problems encountered, 3) monitoring of internal PLC faults, 4) monitoring of PLC output faults, 5) monitoring of power supplies, 6) strict control of program access, and 7) configuration control of documentation.

Stocking of critical spare parts. The stocking of spare parts is a very important maintenance practice, because it reduces down time in the event of a component failure.

Documentation of problems encountered. Detailed maintenance records should be maintained of all faults and problems encountered. The records should include a code of the problem type, a detailed description of the problem encountered, a description of the corrective action taken, and a list of the part or parts replaced or repaired to correct the problem.

Monitoring of internal PLC faults. Most PLC programming packages include diagnostic software that monitors for internal software faults. If a fault or faults are detected, internal bits are set. These bits can be used to alert the user of the problem and the fault bits can be used in the control program to halt operations if the problem is serious.

Monitoring of PLC output faults. By using additional wiring, other inputs, and software logic, it is possible to ensure that critical outputs react in an appropriate way. Alarms or safety relays can also be activated if, for example, safety related outputs fail to operate properly.

Monitoring of external power supplies. The consequences of the loss of power supplies or power sources on the control system should be studied. In some cases, the input or output circuit, either completely or partially, are powered by different electrical sources than the source supplying the PLC processor. These sources should be monitored and the consequences of their loss to the control system should be clearly defined.

Strict control of program access. In order to avoid problems caused by unauthorized program modifications, it is recommended that access to programs be limited and controlled with passwords and security locks.

Updating of documentation. After each control program change is made, installed, and tested, the software should be saved and a new printout generated. It is important to maintain an easily accessible and understandable printout of the control program. Backup copies of program diskettes should be made and stored in a different location. A written description of the program changes should be made so the program's evolution from it's original installation to present configuration can be tracked in case problems occur later.

Troubleshooting

Troubleshooting can be defined as the methods used to determine why a system or component is not functioning properly. As with many practical skills, troubleshooting is an art as well as a practice requiring analytical or logical skill. As such, troubleshooting procedures are a trainable skill. Figure 12-3 shows a flow diagram for the basic analytical methodology that should be used to solve any problem.

The first step in troubleshooting is to identify the problem. Typically, this consists of describing the problem by listing the symptoms and possible solutions. The next step is to collect information about the problem. This data gathering includes questioning the person who reported the problem for more detailed information, viewing physical symptoms, and reviewing the components and systems involved.

The third step is to narrow the problem down. If it involves complex, interactive components, try to narrow it down to the specific component causing the problem. Once the problem has been identified, apply the appropriate troubleshooting method or methods to solve it. If you are unsuccessful, return to the appropriate step in the process and start again.

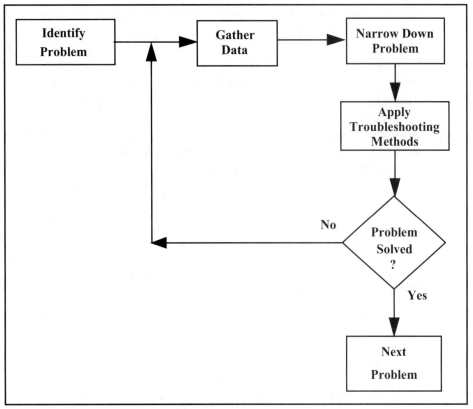

Figure 12-3. Troubleshooting flowchart.

Troubleshooting Methods

There are many different methods of troubleshooting, and each has its advantages and disadvantages. The approach maintenance personnel choose is often based on personal preference or convenience, but in some cases it is dictated by the problem itself. In most cases, more than one method will be used or sometimes even intermixed.

We will discuss the following six main troubleshooting methods: experience, process of elimination, divide and conquer, remove and conquer, substitution, and setting a trap.

Experience

Using your experience is the most common troubleshooting method, and many times it is the simplest. You know the problem and its solution because you have seen it or heard of it before. If there are a number of possible solutions, you can use your experience to determine the area to attack first. Which option you select can be based on highest probability, the one that poses the least risk or upset, the easiest, the closest, the most easily testable, and so on.

Experience is primarily a skill learned on the job: the more work you've done, the more experience you have. Experience can also be gained through hands-on laboratory and classroom training, but if this training is not reinforced, it may soon be forgotten. Experience can be greatly enhanced if the troubleshooter develops the skill to extrapolate his or her experience to a wider range of problems. This method can sometimes have the disadvantage that the troubleshooter will know what is causing the symptoms but not necessarily why the problem is occurring.

Experience can also be formalized by keeping good maintenance records, but this requires that you also have the ability to search and find the information you need.

Process of Elimination

This troubleshooting method is a simple question-and-answer procedure. It starts by asking the following questions:

- What is the problem, and what is it not?
- Where is the problem, and where is it not?
- When does the problem occur, and when does it not?
- What has changed, and what has not?

This technique uses observation, experience, and testing to narrow down the problem into a more workable form. This method is not always linear; that is, the next troubleshooting step may not depend on the prior step.

Divide and Conquer

This method is commonly taught in technical schools and courses. It consists of dividing or breaking the system down into two parts and then testing and finding out which part is working properly and which is not. The part that is not working is further divided and tested again until the cause of the problem is obtained, as shown in Figure 12-4.

The secret to this method is knowing where to divide the system. The dividing point may be based on your experience, its ease of access, test points, and the like. When in doubt, divide it in half. This method works well in PLC systems because you can easily divide the system components into field devices, I/O modules, CPU, or control software.

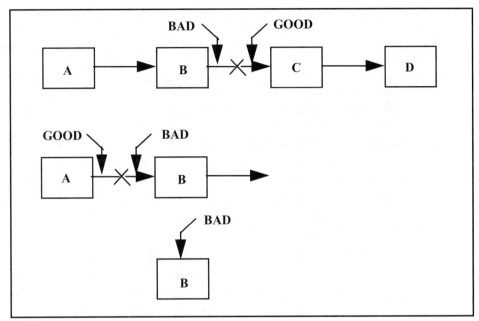

Figure 12-4. Divide-and-conquer troubleshooting method.

Remove and Conquer

This method works best with loosely coupled systems. An example might be where several programmable logic controllers (PLCs) are connected on a communications network. If data is being corrupted, removing each PLC in turn may help you determine if one of the PLCs is causing the problem.

In complex systems, many discrete modules and electronic circuit cards are called on to do different or similar functions. A method for solving problems in these types of systems is to remove the modules or cards from the system one at a time and then see if the problem goes away. Alternatively, remove all the modules or cards, and add them back in until the problem comes back. This method is very effective in modular PLC I/O systems. A block diagram illustrating this method is given in Figure 12-5.

Substitution

This troubleshooting method consists of substituting a known good component into the system to see if the problem goes away. This is typically useful on complex, black-box systems. Experience and testing can limit the number of boxes you substitute.

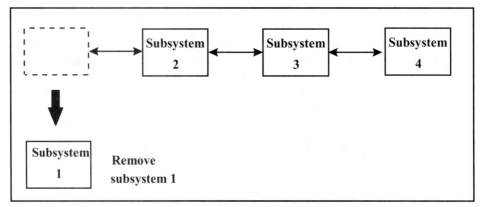

Figure 12-5. Remove-and-conquer troubleshooting method.

A good conceptual understanding of the system that you are troubleshooting is critical to the success of this method. It is also the method of last resort for subtle or intermittent problems. A block diagram illustrating this troubleshooting method is given in Figure 12-6.

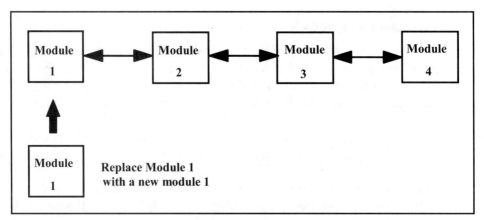

Figure 12-6. Substitution troubleshooting method.

Setting a Trap

Setting a trap is often the only effective method for catching the cause of a spurious or transient problem. With the widespread use of advanced PLC systems that include data logging, archiving, and trending features, much more data is available than in the past. Even so, this data sometimes is not enough, and additional monitoring points or traps are needed to catch the ``prey'' in question. This method is very effective in software troubleshooting where the program can be halted if a certain condition or conditions are detected.

Documentation

Proper documentation of your control system's hardware and software will reduce its design, fabrication, testing, operating, and maintenance costs. Documenting complex control systems generally consists of the following elements: (1) software program listings, (2) backup copies of control programs and configuration software, (3) operations and maintenance (O&M) manuals, (4) system architecture drawings, (5) input/output wiring diagrams, (6) instrument loop diagrams, (7) process and instrument drawings (P&IDs), (8) equipment layout drawings, and (9) mechanical flow diagrams (MFDs).

The system design drawings should contain a detailed list of the components in the system so the maintenance department can order and stock replacement parts. The parts list should include the item number cross-referenced to the component on the drawing, a detailed description of the part, the manufacturer name, and the part number.

Your facility should have a configuration control system that maintains all drawings and documentation in an "As-Built" state. The first step is to verify that all documentation for a control system is accurate and complete during the start-up and commissioning of the system. After the system is placed in use, the configuration control system must keep track of all changes and update the documentation on the installed systems accordingly.

EXERCISES

12.1   Discuss the purpose and advantages of placing programmable controller components in metal enclosures.

12.2   Discuss the factors behind the need to design control panels to meet the heat requirements of programmable controller system components.

12.3   List some important design practices for reducing the maintenance costs on programmable controller systems.

12.4   What are the proper grounding techniques for programmable controller design?

12.5   List the phases in the operational test procedure for a typical control system.

12.6   What are the important features of an effective preventive maintenance program for programmable controller systems?

12.7   Name seven recommended maintenance measures that should be used on PLC control systems.

12.8   What are the troubleshooting methods most commonly used to find and correct problems in PLC control systems?

12.9   List the typical types of documentation used on complex control systems.

12.10   Design a power distribution system for the two-tower dehydration PLC control system discussed in chapter 11. Assume that the system is mounted in a control panel with a six-socket ac outlet power strip, two cooling fans, and a single fluorescent light.

BIBLIOGRAPHY

1.   Allen-Bradley Co. Inc. *Assembly and Installation Manual: PLC5 Family Programmable Controllers*, Publication 1785-6.6.1 (Allen-Bradley, November 1987).

2.   Allen-Bradley Co. Inc. *Assembly and Installation Manual: PLC-2/20, PLC-2/30 Programmable Controllers,* Publication 1772-6.6.2 (Allen-Bradley, March 1984).

3.   Bryan, L. A., and E. A. Bryan. *Programmable Controllers: Theory and Implementation* (Industrial Text Co., 1988).

4.   National Electrical Manufacturers Association (NEMA). *Enclosures for Electrical Equipment (1000 Volts Maximum)*, NEMA Standard Publication No. 250 (NEMA, 1979).

Appendix A
Answers to Exercises

Chapter 1

1.1　The processor reads the inputs, executes logic as determined by the application program, performs calculations, and controls the outputs accordingly.

1.2　The main purpose of the I/O system is to provide the physical connection between the process equipment and the PLC system

1.3　Typical discrete input devices found in process industries include process switches for temperature, flow, and pressure.

1.4　Some typical discrete output signals encountered in industrial applications are control relays, solenoid valves, and motor starters.

1.5　Some typical analog signal values found in process control applications are 4-to-20-ma dc current, 0 to 10 vdc, and 1 to 5 vdc.

1.6　The difference between volatile and nonvolatile memory is that volatile memory will lose its programmed contents if all operating power is lost or removed. Nonvolatile memory will retain its data and program even if there is a complete loss of operating power.

1.7　The common applications for personal computers in programmable controller systems are programming, graphical user interfaces, and data collection.

1.8　The personal computer is the device most commonly used to program PLCs.

1.9　The three basic instructions used in ladder logic programs are normally open, normally closed, and coil output.

1.10   Power flow in a relay ladder diagram is the flow of electric current from the left side to the right side of a ladder rung.

1.11   Logical power flow is obtained in a LAD program if there is logical continuity from the left side to the right side of a ladder logic rung.

1.12   STL instruction statements have two basic structures: a statement made up of an instruction alone (for example, NOT) and a statement made up of an instruction and an address.

Chapter 2

2.1 $1010_2 = 10_{10}$.

2.2 $10011_2 = 19_{10}$.

2.3 $10101011_2 = 253_8$.

2.4 $101111101000_2 = 5750_8$.

2.5 $33_{10} = 41_8$.

2.6 $451_{10} = 703_8$.

2.7 $47_{16} = 71_{10}$.

2.8 $157_{16} = 343_{10}$.

2.9 $56_{10} = 38_{16}$.

2.10   BCD code is 0011 0111.

2.11   BCD code is 0010 0111 0000.

2.12   The ASCII code is 4C 65 76 65 6C 20 4C 6F 77 in Hex format.

2.13   To verify the logic identity $A + 1 = 1$, we let $B = 1$ in the following two-input OR truth table:

Inputs	Output
A B	Z
0 1	1
1 1	1

Since $B = Z = 1$ in the truth table, then $A + 1 = 1$.

2.14   Let start function = A, the stop function = B and the run request = C, so the logic equation is: $(A + C) \bullet \bar{A} = C$.

2.15   Let tank high level = A, the tank low level = B and the Pump Running = C, so the logic equation is: $(\bar{A} \bullet C) \bullet (B + C) = C$.

2.16 The logic gate circuit for a standard start/stop circuit is shown in Figure E2-16.

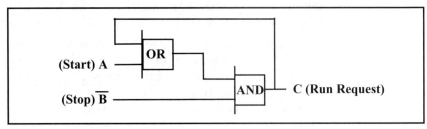

Figure E2-16. Answer to Exercise 2.16.

Chapter 3

3.1 The current flow is 15 amps.

3.2 The current is 4.8 amps.

3.3 The area is 9 cmil.

3.4 The resistance of 500 feet of copper wire with a cross-sectional area of 4,110 cmil is 1.26 Ω.

3.5 The resistance of 500 feet of silver wire with a diameter of 0.001 inch is 4,900 Ω.

3.6 The distance to the point where the wire is shorted to ground is 1,238.2 ft.

3.7 The total current (I_t) in a series circuit with two 250 Ω resistors and an applied voltage is 24 vdc is 48 mA. Also, $V_1 = 12$ volts and $V_2 = 12$ volts.

3.8 $I_1 = 1$ A, $I_2 = 0.5$ A, and $I_t = 1.5$ A.

3.9 $I_1 = 1$ mA, $I_2 = 5$ mA, $R_x = 400$ W, $V_1 = 2$ volts, and $V_2 = 10$ volts.

3.10 Electric relays operate as follows: when a changing electric current is applied, a strong magnetic field is produced, and the resulting magnetic force moves the iron core that is connected to a set of contacts. These contacts in the relay are used to make or break electrical connections in control circuits.

3.11 The three common types of solenoid valves are direct acting, internal pilot-operated, and manual reset.

Chapter 4

4.1   The input/output (I/O) system provides the physical connection between the process equipment and the PLC. It uses various interface circuits to convert the sensed or measured physical quantities of the process, such as motion, level, temperature, pressure, and position, into a logic signal the PLC processor can use. Based on the status of the sensed or measured values, the control program in the PLC processor uses various output circuits or modules to activate devices, such as valves, motors, pumps, and alarms, to exercise control over a machine or process.

4.2   The backplane on a PLC I/O rack is a printed circuit card that contains the parallel communications lines to the processor and the dc voltages that are required to power the logic and interface circuits in the I/O modules.

4.3   Discrete inputs signals are signals from field devices like ON/OFF or OPEN/CLOSED. Five typical examples of discrete input signal devices are pushbuttons, pressure switches, relay contacts, level switches, and temperature switches.

4.4   Five examples of typical discrete output field devices are alarm lights, electric control relays, motor starters, electric valves, and alarm horns.

4.5   The purpose of the bridge rectifier in the ac input circuit shown in Figure 4-2 is to convert the ac input voltage into a dc voltage signal to be used by the PLC.

4.6   The purpose of the optical coupler in the ac input circuit shown in Figure 4-2 is to provide the electrical isolation between the input ac voltage and the dc voltage used by the logic section of the PLC.

4.7   The operation of the typical ac output module shown in Figure 4-5 is as follows: first, the processor sends out a logic signal of 0 or 1 to the logic section of the ac output circuit. The signal from the logic section is then passed through an isolation circuit. The logic signal from the isolation circuit is finally fed to an ac power switch circuit and a filter section to control the ac-operated field device.

4.8   The wiring diagram for Exercise 4.8 is shown in Figure E4.8.

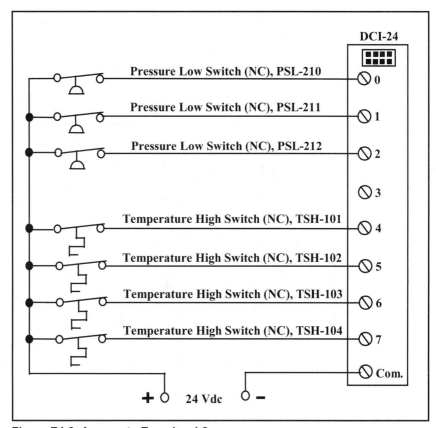

Figure E4-8. Answer to Exercise 4.8.

4.9   Figure E4.9 shows the wiring diagram for a typical +120 vac output
        module connected to four solenoid valves (TV-100, 101, 102, and
        103) and three pump starters (P-100, 200 and 300) with overload
        switches.

4.10   The backplane current requirements for a sixteen-slot I/O rack
        with the following modules installed: (a) four 12 vdc input
        modules (model number DCI-12), (b) six 12 vdc output modules
        (model number DCO-12), (c) four analog input modules (model
        number AI-4-20MA), and (d) two analog output modules (model
        number AO-4-20MA) are calculated as follows:

                    DCI-12: 200 mA,
                    DCO-12: 200 mA,
                    AI-4-20MA: 400 mA, and
                    AO-4-20MA: 400 mA.

So that, the total backplane current required is

4×200 mA = 800 mA
6×200 mA = 1200 mA
4×400 mA = 1600 mA
2×400 mA = $\underline{800}$ mA
$\phantom{2 \times 400 \text{ mA} = } 4400$ mA

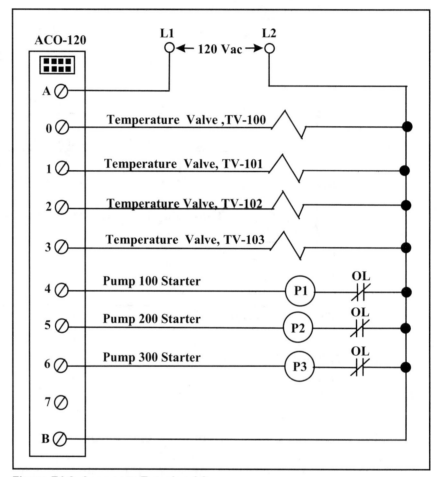

Figure E4-9. Answer to Exercise 4.9.

Chapter 5

5.1   The memory types are read/write memory (R/W), read-only memory (ROM), programmable read-only memory (PROM), and electrically erasable programmable read-only memory (EEPRROM).

5.2   Read/write (R/W) memory is designed so data or information can be written into or read from any unique location. Data or programs are placed into R/W memory using the write mode, and they can be retrieved from R/W memory using the read mode. A typical application for R/W memory is to store a PLC control program.

5.3   The main purpose of read-only memory in PLC applications is to store the executive or operating system program of the PLC.

5.4   There are two main parts of a typical programmable logic controller memory: the system memory and the application memory. The *system memory* is a permanently stored collection of programs and registers that consists of the operating system program, diagnostics software, and system status registers. The operating system directs activities such as executing the control program, communicating with the peripheral devices, and other system housekeeping functions. The *application memory* consists of the input area, output area, data or information storage registers, internal storage bit area, and the control program.

5.5   The input and output image tables store the I/O bits that are used by the PLC control program.

5.6   The memory size required for a programmable controller system with 235 input points and 195 outputs and assuming 25 percent spare memory capacity is 10,750 words.

5.7   The memory bit address for a discrete field device connected to an Allen-Bradley PLC5 input module in rack 1 at I/O group 3 and terminal 1 is I:013/01.

5.8   The memory bit address for a discrete field device connected to an Allen-Bradley PLC5 output module in rack 2 at module group 1 and terminal 7 is O:021/07.

5.9   The memory bit address for a discrete field device connected to an Allen-Bradley PLC5 output module in rack 2 at I/O group 2 and terminal 13 is O:022/13.

5.10   The bit addresses available for use by a sixteen-bit discrete output module installed in slot 4 of rack 1 are Q32.0 through Q32.7 and Q33.0 through Q33.7.

5.11   The bit addresses available for use by a sixteen-bit discrete input module installed in slot 5 of rack 2 are I68.0 through I68.7 and I69.0 through I69.7.

Chapter 6

6.1   The ladder logic diagram program to start a pump that fills a process tank with fluid until a high level is reached in the tank is shown in Figure E6-1. The program assumes that the fill tank pushbutton (PB1) is wired to an Allen-Bradley PLC5 discrete input module at bit I:010/00 and that a NC tank high-level switch is connected to input bit I:010/01. The program also assumes that the pump start relay is connected to output O:000/01 on the A-B PLC.

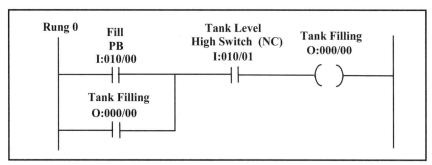

Figure E6-1. LAD program (answer to Exercise 6.1).

6.2   The ladder diagram program to control an electric motor is shown in Figure E6-2. The program assumes that a normally opened start pushbutton is wired to an input module at I:000/00 and a that normally closed stop pushbutton is connected to the same input module at address I:000/01. The program also assumes that PLC5 output O:001/03 is connected to the motor starter and that the normally opened auxiliary contacts on the motor start contactor are connected to PLC input I:000/02.

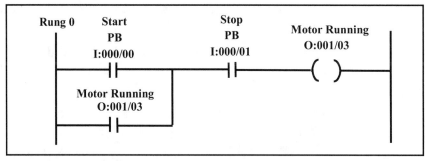

Figure E6-2. Start/Stop LAD program (answer to Exercise 6.2).

6.3   The ladder logic program to control a process pump is shown in Figure E6-3. The program assumes the following conditions: the pump is turned on five seconds after both the inlet and outlet valves to the pump have been opened and the pump is turned off, if either the inlet or the outlet valves are closed. The program also assumes that a pump starter is connected to output bit O:001/00, that a valve-open position switch on the inlet valve is wired to input I:000/02, and that a valve-open position switch on the outlet valve is wired to input I:000/03.

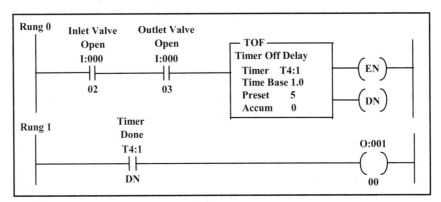

Figure E6-3. Answer to Exercise 6.3

6.4   The LAD program to open the outlet valve (LV-1) on process tank is shown in Figure E6-4. The program assumes that valve LV-1 will be opened if the normally opened level switch (LSH-1) closes when the automatic position of the HOA (HS-1) is selected or if the operator places the HOA switch in the hand position. The program also assumes that the high-level switch is connected to input I:002/01, that the output valve is connected to output O:003/01, that the "Auto" position of the HOA switch is connected to input I:002/02, and that the "Hand" position of the switch is wired to input I:002/03 on a A-B PLC5.

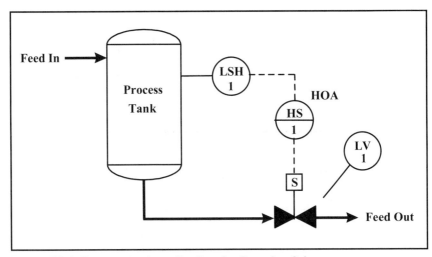

Figure E6-4. Process tank application for Exercise 6.4.

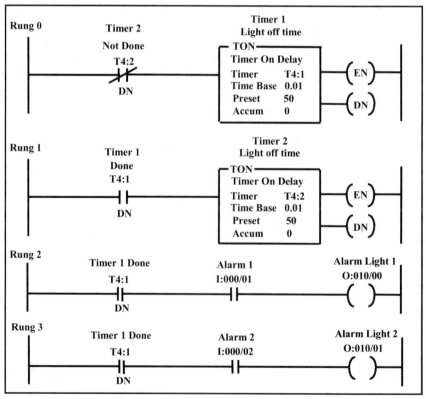

Figure E6-4. Answer to Exercise 6.4.

6.5   The LAD program using timer instructions to flash two process alarm lights on a control panel is shown in Figure E6-5. The two alarm lights are driven by PLC output signals O:010/00 (alarm 1) and O:010/01 (alarm 2). The program assumes that alarm 1 is activated by input I:000/01 and that alarm 2 is activated by input I:000/02.

Figure E6-5. Answer to Exercise 6.5.

6.6 The ladder diagram program to control the temperature of the fluid in a process tank close to 400°C is shown in Figure E6-6. The program assumes that the heater contactor is connected to PLC5 output O:001/01, that a temperature-low switch set at 395°C is connected to input I:000/00, and that a temperature-high switch set at 405°C is connected to input I:000/01. The temperature-low switch is closed if the fluid temperature is below 395°C and opened at a temperature of 395°C or higher. The temperature-high switch is closed if the fluid temperature is below 405°C and opened at 405°C or higher temperatures.

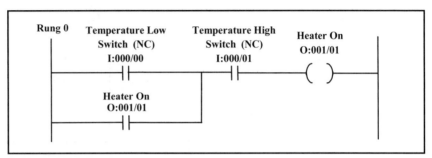

Figure E6-6. Answer to Exercise 6.6.

6.7 Figure E6-7 shows the LAD program for an Allen-Bradley PLC5 that turns off a conveyor belt on a production line after fifty parts have been produced. The program assumes the following: (1) when output bit O:001/12 set to 1 the conveyor belt is turned on; (2) input I:002/01 changes from 0 to 1 and then back to 0 each time a production part is rejected; (3) input I:002/02 changes from 0 to 1 and then back to 0 each time a new part is produced; (4) a normally open (NO) pushbutton connected to input I:002/03 is used to set the production count to fifty, and (5) a NO pushbutton connected to input I:002/04 is used to reset the counter to zero and to stop the conveyor belt.

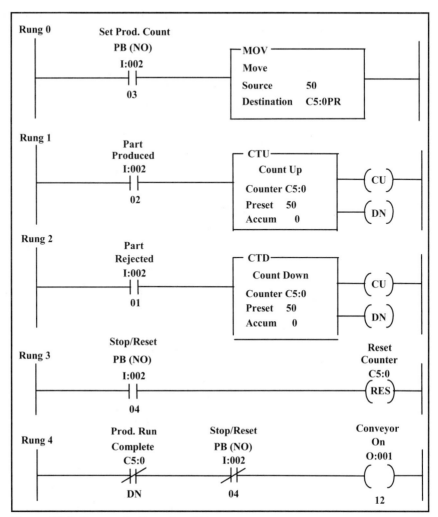

Figure E6-7. Answer to Exercise 6.7.

6.8   Figure E6-8 shows the A-B PLC5 program to add the integer number in word N7:2 to the integer number in word N7:4, to then divide the result stored in N7:20 by 5, and to store the final result in word N7:6.

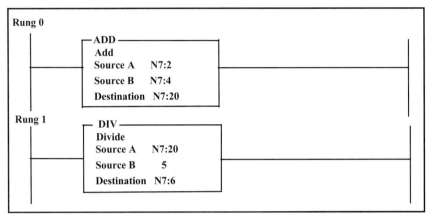

Figure E6-8. Answer to Exercise 6.8.

6.9 Figure E6-9 shows the ladder logic program to transfer the integer number in word N7:0 to an output display at word N7:10 if the number in word N7:0 is greater than the number in word N7:1 and less than the number in location N7:2.

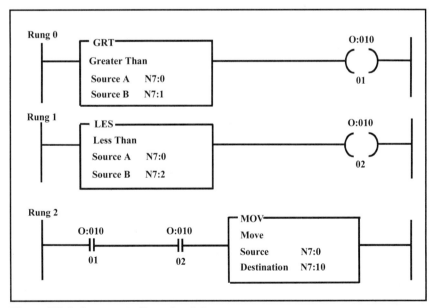

Figure E6-9. Answer to Exercise 6.9.

6.10 Subroutines are used in programming to produce a more structured program and to reduce the amount of memory a program uses. Subroutines are used to store recurring logic functions that can be accessed from different parts of the main ladder logic program. This saves memory space because the function has to be programmed only once but is used many times in the control application.

Chapter 7

7.1   The most common advanced LAD instructions encountered in the Allen-Bradley PLC5 family are file, shift register, sequence and block transfer.

7.2   A file is a group of consecutive data table words used to store information in a PLC system.

7.3   The FAL instruction performs copy, arithmetic, logic, and function operations on the data stored in files. The FAL instruction is an output instruction that performs the operations defined by the source addresses and the operators listed by the programmer in the expression field.

7.4   Figure E7-4 shows the PLC5 LAD program to copy the process data located in integer file #N30, words 0, 1, 2, and 3, to integer file #N31, starting at word 0, if input bit I:001/01 is true.

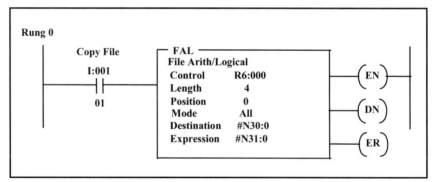

Figure E7-4. Answer to Exercise 7-4.

7.5   Figure E7-5 shows the Allen-Bradley PLC5 LAD program to review the data in integer file #N50, words 0 through 20, and compare it for an equal condition to the data stored in file #N60, starting at word 1.

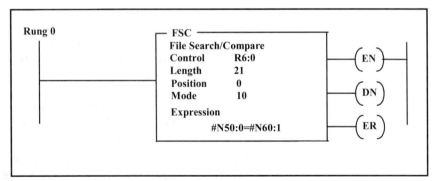

Figure E7-5. Answer to Exercise 7-5.

7.6   The *control* is the address of the control structure in a control type
       (R) file. The processor uses this information to run the instruction.
       The *length* is the number of words (0 to 999) in the data block on
       which the file instruction operates. The *position* is the current
       element within the data block that the processor is accessing. The
       mode is the number of file elements that are operated on each time
       the rung is scanned in the program. There are three modes: All,
       Numerical, and Incremental. In the All mode, the entire file is
       operated on before continuing on to the next rung of the program.
       The Numerical mode distributes the file operation over a number
       of program scans. The Incremental mode manipulates one word of
       the file each time the rung goes from false to true. The *destination* is
       the address at which the processor stores the result of the
       operation. The instruction converts to the data type specified by
       the destination address. The *expression* contains addresses,
       program constants, and operators that specify the source of data
       and the operations to be performed.

7.7   Shift register instructions are used to track the movement or flow
       of parts and information in industrial applications.

7.8   There are three common sequence instructions: sequencer input,
       sequencer output, and sequencer load. They are generally used to
       transfer data from the memory to discrete output modules for
       controlling sequential process operations or sequential batch
       operations (sequencer output). They are also used to compare I/O
       word data with data stored in tables so process operating
       conditions can be examined for control and diagnostic purposes
       (sequencer input). The sequencer load instruction is also used to
       transfer I/O word data into the memory.

7.9   Length is the number of words in a sequencer file.

7.10   Position is the current position of the word in the sequencer file
       that the processor is using.

Chapter 8

8.1   The five PLC programming languages defined in the IEC 1131-3 standard are: (1) ladder diagram, (2) function block diagram, (3) sequential function chart, (4) statement list, and (5) structured text.

8.2   The two text-based PLC languages are statement list and structured text. Statement list, or STL, is a low-level programming language. It is very effective for small simple applications or for optimizing parts of an application. Instructions always relate to the current result, and the operator indicates the operation that must be made between the current value and the operand. The result of the operation is stored again in the current result. Structured text (ST) is a high-level structured language designed for automation processes. This language is used mainly to implement complex procedures that cannot be easily expressed with graphical languages. ST is the default language for the description of the actions within the steps and conditions attached to the transitions of the SFC language.

8.3   The three graphical PLC languages are ladder diagram, function block diagram, and sequential function chart.

The ladder diagram (LAD) is the most common programmable controller language. It consists of a set of instructions that will perform the most basic types of control functions: relay type logic, timing and counting, and basic math operations. However, depending on the programmable controller model used, the instruction set may be extended or enhanced to perform other operations. These additional functions are used for analog control, data manipulation, reporting, complex control logic, and other functions.

The function block diagram (FBD) is a graphical programming language that uses logic blocks similar to those used in Boolean algebra to represent basic logic. It also uses more complex function blocks to perform operations such as timing, counting, math, loading data, transferring data, and data comparison. The programmer is able to build complex control schemes by using functions from the FBD library and then interconnecting them in a graphical diagram area.

Sequential function chart (SFC) is a graphical language used to describe sequential operations. The control process is represented as a set of well-defined steps linked by transitions. A Boolean logic condition is attached to each transition. Actions within the steps and the logic transitions between them can be performed by using instructions from the other standard PLC languages.

8.4   The "initial" step instruction in a sequential function chart (SFC) program is used to start the program.

8.5   The "transition" instruction in an SFC program is the logic condition that the processor checks after completing the active step. When the transition logic is TRUE, the step preceding the transition is disabled, and the step following it becomes active.

8.6   The programming tips that can be used to improve the readability of a structured text (ST) program are as follows: (1) do not write more than one statement on one line, (2) use tabs to indent complex statements, and (3) insert comments to increase the readability of lines.

8.7   STL instruction statements have two basic structures. The first is a statement consisting of an instruction alone (for example, NOT), and the second is a structure in which the statement consists of an instruction and an address.

8.8   The STL program to control the pump is as follows:

        A I124.0
        A I124.2
        = Q124.0

8.9   The five timer instructions available in the Siemens S7 STL programming instruction set are pulse timer (SP), extended pulse timer (SE), on-delay timer (SD), retentive on-delay timer (SS), and off-delay timer (SF).

8.10   The STL program to open the fill valve for thirty seconds is listed below. An extended pulse timer (SE) was used because the pushbutton to open the valve is a momentary type, so the "fill valve" logic signal is present only for a short time.

STL Instruction	Description
A I124.6	Check logic of fill tank input I124.6
L S5T#30S	Load thirty-second time delay if Pump Run bit is 1
SE T3	Start timer T2 as an "extended pulse" timer
A I124.2	Check for bit I124.2 = 1 to reset timer
R T3	Reset timer, if I124.2=1
A T3	Check timer output status
= Q124.2	Open fill valve for thirty seconds

8.11   This is the STL program to control the production line:

STL Instruction	Description
A I124.7	Part rejected
CU C4	Count up
A I124.6	Part produced
CD C4	Count down
A I124.2	Production count input bit check
L C#10	Load production count value
S C4	Set count to 10
A I124.3	Check for bit I124.3 = 1 to reset or stop
R C4	Reset count to zero, if I124.3=1
A C4	Check counter output status
= Q124.5	Run conveyor if count value not zero

8.12   The STL program to perform the math operation is as follows:

STL Instruction	Description
L MW60	Load word MW60 into accumulator 1
L MW62	Load word MW62 into accumulator 1
+I	Add MW60 to MW62
L 10	Load integer 10 into accumulator 1
*I	Divide the result of addition by 10
T MW50	Store result in MW64

8.13   The STL program to perform the math operation is as follows:

STL Instruction	Description
L MD50	Load value in MD70 into Accumulator 1
L MD54	Load value in MD74 into Accumulator 1, CPU transfers MD70 value into Accumulator 2
-R	Subtract MD70 from MD74
L 1.000E+01	Load 10 into Accumulator 1
/R	Divide 10 into result of (MD74 - MD74)
T MD78	Transfer results to MD58

Chapter 9

9.1   The four basic combinations of elements and boxes used in FBD
        programming are (1) instructions as elements, (2) instruction as a
        box with address, (3) instruction as a box with address and value,
        and (4) instruction as a box with parameters.

9.2   The FBD program to control filling a tank is shown in Figure E9-2.
        The program starts a pump that fills a process tank with fluid until
        a high level is reached. The program assumes that a normally
        opened (NO) fill tank pushbutton is wired to an input at I124.4 and
        that a normally closed (NC) tank high-level switch is connected to
        input I124.5 on a Siemens PLC model CPU314IFM. It also assumes
        that the pump start relay is connected to output Q124.0 on the
        same Siemens PLC.

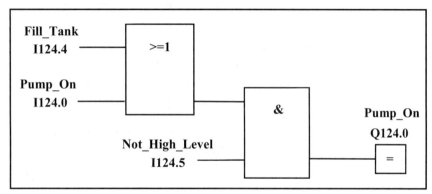

Figure E9-2. Answer to Exercise 9-2.

9.3   The FBD program to control an electric motor is shown in Figure
        E9-3. This program assumes that a NO start pushbutton is wired to
        input I124.0 and that an NC stop pushbutton is wired to input
        I124.5 of a Siemens PLC model CPU314IFM. Also, the program
        assumes that output Q124.1 of the same PLC controls a motor start
        relay and that the auxiliary (NO) contacts of the motor starter are
        connected to the PLC input I124.1.

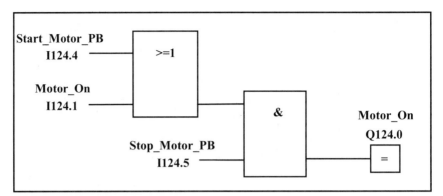

Figure E9-3. Answer to Exercise 9-3.

9.4   The five timer instructions available in the Siemens S7-300 FBD programming instruction set are (1) Pulse, (2) Extended Pulse, (3) On-delay, (4) Retentive On-delay, and (5) Off-delay.

9.5   The FBD program to control the temperature of the fluid in a process tank is shown in Figure E9-5. The program assumes that the heater contactor is connected to output Q124.1, a temperature-low switch is connected to PLC input I124.0, and a temperature-high switch is connected to PLC input I124.1 on a Siemens model CPU314IFM programmable controller. The temperature-low switch is closed if the fluid temperature is below 395°C and opened at a temperature of 395°C or higher. The temperature-high switch is closed if the fluid temperature is below 405°C and opened at 405°C or higher temperatures.

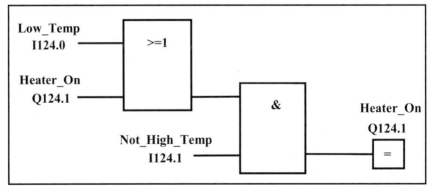

Figure E9-5. Answer to Exercise 9.5.

9.6   The FBD program to control the pump is shown in Figure E9-6. This program assumes that a pump starter relay is connected to output point Q124.0 and that if the inlet control valve FV-1 on a process tank is opened (input bit I124.0 =1) and the process tank is at a high level (input bit I124.2 = 1) the pump will be turned on.

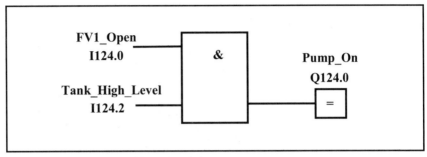

Figure E9-6. Answer to Exercise 9.6.

9.7 Figure E9-7 shows the FBD program to cause a Simatic CPU314IFM PLC to open a fill valve on a process tank to allow an ingredient to be added to the tank for ten seconds each time the Tank Level High Switch LSH-100 is activated. Note that switch LSH-100 is connected to input I124.0 and that the fill valve is wired to PLC output point Q124.0.

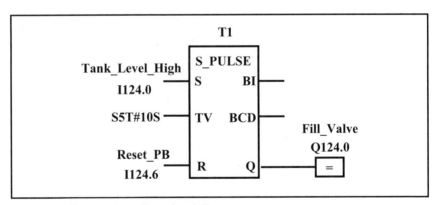

Figure E9-7. Answer to Exercise 9-7.

9.8 The FBD program to control the conveyor belt on a production line is shown in Figure E9-8. The control program turns off the conveyor after fifty parts have been produced. The program assumes the following: (1) Output bit Q124.5 = 0 turns off the conveyor belt; (2) Input I124.1 is set to one (1) each time a production part is rejected; (3) Input I124.2 is set to one (1) each time a new part is produced; (4) A normally open (NO) pushbutton connected to input I124.6 is used to set the production count to fifty, and (5) A NO pushbutton connected to input I124.7 is used to reset the counter to zero and to stop the conveyor belt.

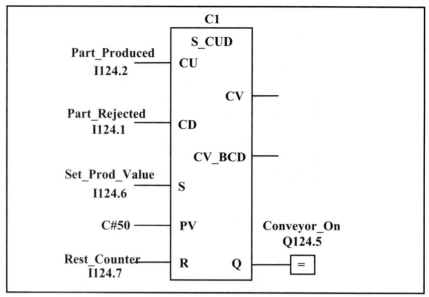

Figure E9-8. Answer to Exercise 9-8.

9.9   The FBD program to perform the math function is shown in Figure
       E9.9. This program subtracts the integer data in word MW20 from
       the integer data in word MW18 and stores the result in word
       MW22, if the input bit I124.0 is one (1). The program then divides
       the result of the subtraction by two and stores the final result in
       word MW24.

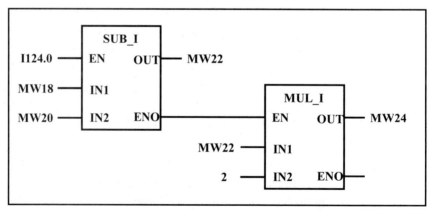

Figure E9-9. Answer to Exercise 9.9.

9.10   The FBD program to perform the math function is shown in Figure
       E9.9. This program subtracts a thirty-two--bit floating-point
       number in word MD70 from a thirty-two-bit floating-point
       number in word MD74 and stores the result in word MD78 if input

I124.0 is set to 1. The program then divides the result by 4.5 and stores the final result in word MD82.

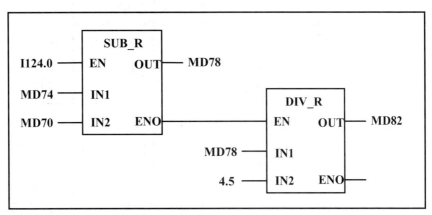

Figure E9-10. Answer to Exercise 9.10.

Chapter 10

10.1   The three basic components of a data communications system are the *transmitter* to generate the information, the *medium* to carry the data, and the *receiver* to detect the data.

10.2   In unidirectional communications, a single channel is used and communication is only one-way (i.e., from transmitter to receiver), so the receiver can never respond. Two-way communications allows the receiver to verify that the data was received. One kind of two-way communications is called half-duplex. In half-duplex, a single channel is used and communication is two-way, but communication can occur only in one direction at a time. Two-way communications in which data can flow in both directions at the same time is called *full-duplex communications*. In this case, two physical paths are required so information can flow from both directions simultaneously.

10.3   In parallel transmission, all bits are transmitted at the same time, and each bit of information requires a unique channel. For example, if we transmit an eight-bit ASCII character, eight channels are required. The term *parallel* refers to the position of the bits of the character and the fact that the characters are transferred as a group of bits in parallel. Using a group of channels results in a high data transfer rate. In serial transmission, the bits of the encoded character are transmitted one after another in a single channel. The transmission takes the form of a bit stream that the receiver must assemble into characters (normally eight bits). The main advantages of serial communications are lower cost and the elimination of byte synchronization problems.

10.4   The three most common methods of signal multiplexing encountered in communications systems are frequency division, time division, and statistical.

10.5   The four most common data transmission error-checking methods used in PLC communications systems are echo check, vertical redundancy check, longitudinal redundancy check, and cyclical redundancy check.

10.6   The LRC/VRC odd-parity characters required for the message M1 ON are given in Table E10-6.

Table E10-6. Answer to Exercise 10.6

ASCII String	M	1		O	N	LRC odd-parity bit
Bit 1	1	1	0	1	0	0
Bit 2	0	0	0	1	1	1
Bit 3	1	0	0	1	1	0
Bit 4	1	0	0	1	1	0
Bit 5	0	1	0	0	0	0
Bit 6	0	1	1	0	0	1
Bit 7	1	0	0	1	1	0
VRC odd-parity bit	1	0	0	0	1	1

10.7   In *asynchronous*, successive data appear in the data channel at arbitrary times with no specific clock control governing the time delays between information. In *synchronous*, each successive datum in a data stream is controlled by a master data clock and appears at a specific interval of time.

10.8   Three common serial data protocols used in communications systems are BISYNC, DDCMP, and HDLC.

10.9   The bit streams required to send the message RUN using eight-bit even-parity ASCII code and BISYNC protocol are as follows: SYNC = 10010110, SYNC = 10010110, SOH = 10000010, STX = 10000010, R = 11010010, U = 110110101, N = 01001110, ETX = 10000011.

10.10   The seven layers of the ISO/OSI communications standard are: (1) physical, (2) data link, (3) network, (4) transport, (5) session, (6) presentation, and (7) application.

Chapter 11

11.1   To design PLC-based control systems you need a process flow
        diagram or equipment layout drawing, a process control
        description, correct sizing and selection of the PLC equipment, a
        system specification, system drawings, wiring diagrams, and
        control programming.

11.2   Piping and instrument drawings show the process equipment that
        is to be controlled and the instrumentation used to control the
        process.

11.3   The function of a process description in PLC system design is to
        convey the purpose and steps of the process being controlled.

11.4   The memory size required for a PLC application with the I/O
        points described is 7K.

11.5   Some of the peripheral equipment commonly used in PLC
        applications are programming devices, mass storage units, and
        printers.

11.6   The purpose of a PLC system drawing is to give an overall view of
        the system component interconnections and the communications
        cabling layout.

11.7   If a third process tower is added, we would need to add four
        solenoid-operated air valves with open and closed limited
        switches. Therefore, we would need to add four ac output points
        or one additional ac output module as well as a single ac input
        module for the eight discrete inputs from the eight-valve position
        switches. We would need a third set of pressure switches (high and
        low) in the application, but the two discrete inputs from these
        pressure switches could be connected to the spare discrete inputs
        on the existing ac input module shown in Figure 11-10.

11.8   The ladder logic program to control the valves on tower 2 of the
        dehydration system application is shown in Figure E11-7.

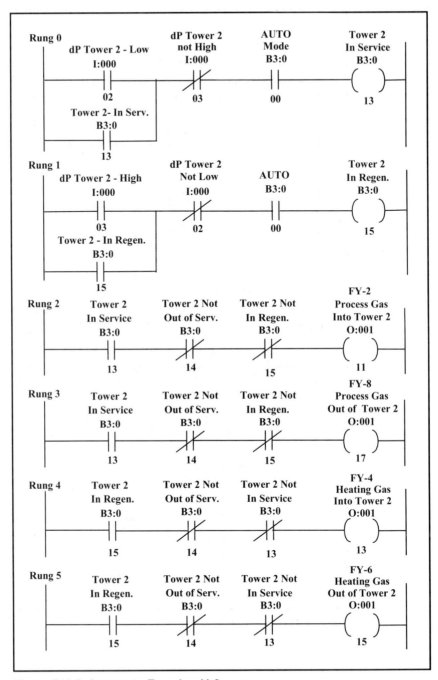

Figure E11-8. Answer to Exercise 11.8.

11.9   Figure E11-9 shows the revision of rungs 2 and 3 of the LAD
        program for alternating pump applications.

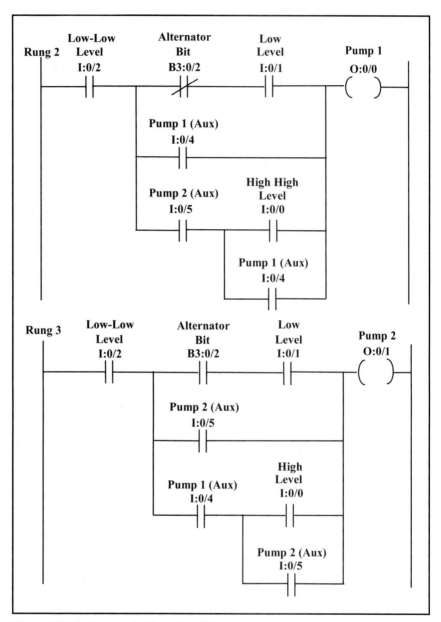

Figure E11-9. Answer to Exercise 11.9.

Chapter 12

12.1   Programmable controller components are placed in metal enclosures to protect against atmospheric contaminants such as dust, moisture, oils, and other corrosive airborne substances. These metal enclosures also reduce the effects of the electromagnetic radiation generated by electrical or welding equipment.

12.2   The temperature inside the control panel or enclosure must not exceed the maximum operating temperature listed in the manufacturer's installation manual (typically 120°F). If that temperature limit cannot be maintained by convection cooling, fans or an air conditioner must be installed to help dissipate the heat. The fans or air conditioner will generally be equipped with filters to prevent dust, dirt, and other airborne contaminants from entering the enclosure and affecting system components.

12.3   The following design features will reduce maintenance cost: (1) proper placement of system components, (2) an ac power outlet strip in the control panel, (3) interior lighting in large panels, and (4) in harsh environmental, a gasketed glass window installed in panels so the processor status lights and/or I/O status indicators can be viewed.

12.4   The National Electric Code states that a ground must be permanent (i.e., no solder connections), continuous, and able to safely conduct the current in the system with minimal resistance. The ground wire should be separated from the ac hot and ac neutral wires at the point at which they enter the control panel. To minimize the length of the ground wire within the panel, the ground reference point should be located as close as possible to the point of entry of the panel supply line. All I/O racks, power supplies, processors, and other electrical devices in the system should be connected to a single ground bus in the control panel. Paint or other nonconductive materials should be scraped away from the area where an I/O rack makes contact with the panel. In addition to making the ground connection through the rack mounting bolts, a metal braid of the size recommended by the PLC manufacturer should be used to connect each chassis and the panel at a single mounting bolt.

12.5   The phases of a typical control system operational test procedure are (1) visual inspection, (2) continuity check, (3) input wiring check, (4) output wiring check, and (5) operational test.

12.6   The main features of a preventive maintenance program for PLC systems are (1) checking all electrical connections on a regular schedule, (2) inspecting all air filters in the system at scheduled intervals, (3) do not move equipment that produces high levels of electrical noise near operating PLC systems, and (4) do not store unnecessary items on or near PLC equipment.

12.7 The recommended maintenance practices are (1) stocking critical spare parts, (2) documenting problems encountered, (3) monitoring internal PLC faults, (4) monitoring PLC output faults, (5) monitoring power supplies, (6) strictly controlling program access, and (7) updating control program documentation.

12.8 The troubleshooting methods most commonly used to find and correct problems in PLC control systems are (1) experience, (2) process of elimination, (3) divide and conquer, (4) remove and conquer, (5) substitution, and (6) setting a trap.

12.9 The typical types of documentation used on complex control systems are (1) software program listings, (2) backup copies of control programs and configuration software, (3) operations and maintenance (O&M) manuals, (4) system architecture drawings, (5) input/output wiring diagrams, (6) instrument loop diagrams, (7) process and instrument drawings (P&IDs), (8) equipment layout drawings, and (9) mechanical flow diagrams (MFDs).

12.10 The power distribution system drawing for the two-tower dehydration PLC control system is shown in Figure E12-10.

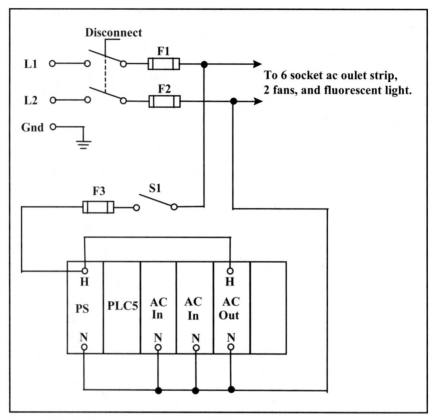

Figure E12-10. Answer to Exercise 12.10.

Index